## Unless Recalled Earlier

### Date Due

| | | |
|---|---|---|
| NOV 2 4 1987 | | |
| DEC 1 5 1987 | | |
| APR 8 1988 | | |
| MAR – 5 2000 | | |
| MAR 2 6 2000 | | |
| | | |
| | | |
| | | |
| | | |
| | | |
| | | |
| | | |
| | | |
| | | |
| | | |

# ENGINEERS
## &
# ELECTRONS

# "WHAT WILL HE GROW TO?"

*A cartoon from the 1881 issue of* Punch, *showing King Steam and King Coal anxiously watching the infant Electricity and asking, "What will he grow to?"*

# ENGINEERS
# &
# ELECTRONS
## A Century of Electrical Progress

*John D. Ryder*
*Donald G. Fink*

**IEEE PRESS**

**The Institute of Electrical and Electronics Engineers, Inc., New York**

Copyright © 1984 by

THE INSTITUTE OF ELECTRICAL
AND ELECTRONICS ENGINEERS, INC.
345 East 47th Street
New York, NY   10017
All rights reserved.

PRINTED IN THE UNITED STATES OF AMERICA
Second Printing

IEEE Order Number PC01669

Library of Congress Cataloging in Publication Data:

Ryder, John Douglas, 1907–
   Engineers and electrons.

   Includes index.
   1. Electric engineering — United States — History.
I. Fink, Donald G.    II. Title.
TK23.R9   1984      621.3′0973      83-22681
ISBN 0-87942-172-X

**A CENTURY OF ELECTRICAL PROGRESS**

## CENTENNIAL TASK FORCE

John D. Ryder, *Chairman*
Donald S. Brereton, *Vice Chairman*

| | |
|---|---|
| Nathan Cohn | Donald T. Michael |
| Robert F. Cotellessa | William W. Middleton |
| Lawrence P. Grayson | Mac E. Van Valkenburg |

## HONORARY CENTENNIAL COMMITTEE

Richard J. Gowen, *1984 IEEE President*
James B. Owens, *Vice Chairman*
Charles A. Eldon, *Vice Chairman*

| | |
|---|---|
| John Bardeen | Ian M. Ross |
| William R. Hewlett | Roland W. Schmitt |
| William C. Norris | Mark Shepherd, Jr. |
| Robert N. Noyce | John W. Simpson |
| Simon Ramo | Charles H. Townes |

# TABLE
# OF
# CONTENTS

Preface . . . xiii

Foreword . . . xv

CHAPTER

## 1

## IN THE BEGINNING
**PAGE 1**

CHAPTER

## 2

## DISCOVERY
## AND INVENTION
**PAGE 11**

Dots and Dashes . . . 11

To Speak and Hear . . . 16

The Genie's Lamp . . . 20

Electric Light for the World . . . 22

For Further Reading . . . 25

CHAPTER

## 3

## FROM
## ELECTRICIANS
## TO ENGINEERS
**PAGE 27**

The Magic Years . . . 27

Organization of the AIEE . . . 28

The Men of the AIEE . . . 29

The International Electrical
Exhibition . . . 31

The First AIEE Meeting . . . 33

The Era of the Central Station
Begins . . . 35

AC Versus DC . . . 35

AC Is Paramount . . . 37

Niagara Falls — An Engineering
Triumph . . . 39

The Electricians Pass . . . 40

Education of the Engineers . . . 41

Standards . . . 43

An Institute Badge . . . 44

The Members Stir . . . 44

For Further Reading . . . 45

CHAPTER

## 4

## MAXWELL'S
## PROPHECY
## FULFILLED
**PAGE 47**

Marconi Takes the Stage . . . 47

The Electron Is Identified . . . 48

The deForest Triode . . . 51

Industrial Research Begins at General Electric . . . 52

Bell System Research . . . 54

Armstrong's First Invention . . . 58

Radio in Chaos . . . 59

Armstrong's Second Invention: The Superheterodyne . . . 60

The Greeks Had a Word for It . . . 61

Meanwhile on the Society Front . . . 61

The Engineering Societies Building . . . 62

Professionalism in the AIEE . . . 63

Membership Requirements . . . 64

Organization of the IRE . . . 65

For Further Reading . . . 67

CHAPTER

# 5

# THE ELECTRON SINGS A GOLDEN TUNE
**PAGE 69**

The Marconi Monopoly . . . 69

RCA Is Organized . . . 70

The Sarnoff Saga . . . 72

Broadcasting . . . 74

Three-Handed Americans: The Neutrodyne . . . 75

200 Meters and Down . . . 76

Armstrong and Superregeneration . . . 77

Radio Enters the Living Room . . . 78

Antitrust Actions . . . 80

Armstrong and Frequency Modulation . . . 80

Armstrong Battles for His Life — and Loses . . . 81

Half a Signal and More . . . 83

Musings on a Ferryboat . . . 83

Bell Meets the Communication Needs . . . 84

More Music in the Home . . . 85

War of the Disks . . . 86

Wire and Tape . . . 87

Pictures Speak . . . 88

Six Decades . . . 89

For Further Reading . . . 89

CHAPTER

# 6

# THE ELECTRON SHOWS ITS MUSCLE
**PAGE 91**

Early Central Stations . . . 91

Developments in Europe . . . 94

Electric Transit . . . 97

Plants and Generators Grow . . . 99

Transmission Line Problems . . . 100

Building a Utility System . . . 103

Theories of Management . . . 105

A Capital-Intensive Industry . . . 106

Government Regulation or Control? . . . 107

Hydroelectric Power . . . 108

Government: Regulator or Competitor? . . . 109

The TVA Dilemma . . . 111

Coal Loses a Market . . . 112

The Nuclear Age . . . 113

Electric Power Research
Institute . . . 114

For Further Reading . . . 115

CHAPTER

# 7

# ELECTRONS AND HOLES: MUCH FROM LITTLE
### PAGE 117

A Name for a New Field . . . 117

Semiconductors: Key to Solid-State
Electronics . . . 118

Developments During World
War II . . . 119

Why Transistors? . . . 120

The Three Inventors . . . 121

The Minority Movement in
Semiconductors . . . 122

The Point-Contact Transistor Is
Invented . . . 122

The P-N Junction: Key to the Junction
Transistor . . . 124

The Junction Transistor . . . 125

Achievements of the Junction
Transistor . . . 126

Changing Forms of the Junction
Transistor . . . 126

Transistors, Et Cetera, on a
Chip . . . 127

How Integrated Circuits Are
Made . . . 129

The Many Uses of Solid-State
Electronics . . . 130

For Further Reading . . . 131

CHAPTER

# 8

# ELECTRONS IN WAR AND PEACE
### PAGE 133

War Stimulates Invention . . . 133

Radar Began with Hertz . . . 134

Microwave Radar . . . 137

Countermeasures . . . 139

Termination of the Work . . . 140

Spin-Offs in Peace Time . . . 140

Sonar: Underwater Radar . . . 141

Electronic Aids to Navigation . . . 142

Mobile Communications . . . 143

From Missiles to Satellites . . . 144

Satellites That Talk Back . . . 145

Eyes in the Skies . . . 146

For Further Reading . . . 147

CHAPTER

# 9

# ALL THE WORLD BECOMES A STAGE
### PAGE 149

Basic Schemes . . . 149

Early Ideas . . . 150

The Nipkow Disk . . . 151

First Demonstrations . . . 153

And in Color, Too! . . . 153

The End of Mechanical
Systems . . . 154

The Dawn of Electronic
Television . . . 155

All-Electronic Television
Arrives . . . 158

Sarnoff and Farnsworth . . . 160

Standards Again . . . 161

The First NTSC . . . 162

Incompatibility in Color . . . 164

Compatibility in Color . . . 164

NTSC II . . . 165

Final Details of the Color
System . . . 167

A Picture Tube for Color . . . 167

Success for the NTSC . . . 168

Progress in Europe and Japan . . . 168

The Role of the Professional
Society . . . 170

The Impact of TV on Society . . . 170

For Further Reading . . . 171

CHAPTER

## 10

# COMPUTERS AND THE INFORMATION REVOLUTION
**PAGE 173**

Information and Its Symbols . . . 173

Codes for the Computer . . . 174

Channels of Communication . . . 175

Babbage and the Computer . . . 177

Babbage and the Countess . . . 179

The Route to the Electronic
Computer . . . 180

COLOSSUS Versus ENIGMA — The
Code War . . . 181

ENIAC . . . 182

Time and Progress . . . 183

Computer Capabilities . . . 184

What Computers Do . . . 185

Programmers and Computer
Languages . . . 186

The Microprocessor . . . 187

Signal Conversion . . . 187

Games People Play . . . 188

Computers and the Human
Mind . . . 188

For Further Reading . . . 189

CHAPTER

## 11

# YOUNG ENGINEERS AND THEIR ELDERS
**PAGE 191**

Liberal and Practical
Education . . . 192

The Roots of Electrical
Engineering . . . 194

The AIEE and Education . . . 194

Conflicting Industry Views . . . 196

Education at a Standstill . . . 198

Development of Graduate
Study . . . 200

Effects of World War II . . . 201

And the Field Moves On . . . 203

Measuring Engineering
Education . . . 203

Community Colleges and Technical
Institutes . . . 204

Continuing Education . . . 205

Contributions of Engineering
Faculty . . . 206

Broadening of the Field . . . 206

For Further Reading . . . 207

C H A P T E R
# 12
# AIEE + IRE = IEEE
**PAGE 209**

*Way Back in 1912 . . . 210*

*Common Objectives — Different Routes . . . 211*

*Two Routes in Publications . . . 212*

*The IRE "Show" . . . 214*

*Philosophies for the Membership Grades . . . 214*

*Differing Viewpoints on Internationality . . . 215*

*The 1947–1962 Membership Race . . . 215*

*The IRE Group Plan . . . 216*

*The Students Are Heard From . . . 217*

*The Membership Race at the Grass Roots . . . 218*

*The AIEE Reaction . . . 219*

*The Merger Looms . . . 220*

*How It Was Done . . . 222*

*The Name and Badge . . . 222*

*The First General Manager . . . 224*

*The IEEE Is Born . . . 225*

*The Birth and Growth of IEEE SPECTRUM . . . 227*

*The IEEE Goes Professional . . . 229*

*Professionalism at Work . . . 230*

*The Centennial Approaches . . . 231*

*For Further Reading . . . 231*

Appendixes . . . 232

Index . . . 240

Picture Credits . . . 249

Authors' Biographies . . . 250

# PREFACE

In 1975 one of the authors of this book, as Chairman of the IEEE History Committee, proposed that a need existed for more research in and publication of the history of the electrical field and profession. The help of several professional historians was sought, and at their suggestion the annual IEEE History Fellowship was established to support a student working on a dissertation in the history of electrical science and technology; financing of the fellowship program was provided by the IEEE Life Member Fund. This program has produced a number of capable historians who are now adding to the historical literature on the electrical field.

A further effort to increase the amount of history publication led to the 1976 IEEE PRESS book, *Turning Points in American Electrical History,* capably edited by an historian, Dr. James Brittain of the Georgia Institute of Technology, and the September 1976 Bicentennial issue of the PROCEEDINGS OF THE IEEE entitled "Two Centuries in Retrospect." The latter, containing papers by noted engineers as well as papers by historians of technology, ably presented the viewpoint that history involves people as well as technology.

By 1979 these programs had provided a wealth of evidence that the electrical field was rich in unrecorded history and the IEEE Board of Directors established the Center for the History of Electrical Engineering at IEEE Headquarters in New York City.

All these efforts in the field of history established an atmosphere in which the 1984 Centennial of the IEEE could be approached. Preparations for a suitable Centennial celebration began in 1978 with the appointment of an Honorary Centennial Committee and a Centennial Task Force to plan the details of a year-long celebration in 1984.

The inspiration for this book came in a conversation with Donald Christiansen, Editor of IEEE SPECTRUM, who suggested that a popularly oriented and illustrated history of electrical engineering and of the IEEE might be a useful contribution to the Centennial celebration. This concept was proposed to and accepted by the Centennial Task Force.

In early 1982 our author team was formed, the book was outlined, a general format and style was agreed upon, and our individual areas were explored for suitable material. Needless to say, we changed the outline several times as the writing progressed during 1982 and early 1983, using our personal knowledge of events and personalities in the field as well as the usual historical sources.

An appeal was made to the IEEE Life Member Fund for financial support to produce an attractive book and to augment the text liberally with illustrations of historic artifacts and people. Dr. Joseph Keithley, Chairman of the Life Member Fund Committee, presented our request to that group, which agreed to supply some of their funds which had been husbanded for Centennial projects. The results of the use of that money appear throughout this volume.

The book represents our efforts to tell, in an informal and readable way, the often exciting story of how electrical engineering began unpretentiously about a century ago, grew impressively in the decades that followed, and yielded the technology that today reaches into all aspects of daily life. The book describes both technical developments and the people behind them; we do not claim that it is comprehensive nor necessarily balanced in the events and people chosen for coverage. Although it does not purport to be a scholarly history, it is our hope that the book may satisfy the desires of the professional for concise information and may thrill engineering students when they learn something about "double E" beyond that gained in their everyday course work. And if other young people are aided in making a correct career decision, our reward will be great.

Most chapters conclude with a list of articles and books for further reading. In addition, the Centennial Task Force has arranged for the publication in early 1984 of a more comprehensive social history of electrical engineering written by a professional historian, Dr. Michal McMahon, entitled *The Making of a Profession: A Century of Electrical Engineering in America*.

The authors wish to acknowledge aid, particularly with illustrations, from Robert Friedel and the IEEE Center for the History of Electrical Engineering, and from W. Reed Crone of the IEEE PRESS, whose efforts and interest have gone well beyond his normal duties.

John D. Ryder
Donald G. Fink

# FOREWORD

In touring the pages of this volume the reader will be reminded that practitioners and professors of electrical engineering and the related sciences, although sometimes considered remote and antisocial, are in fact very human. Here you will meet both the geniuses and the craftsmen who, during the first century of their profession, collectively changed the life-styles of the citizens of our industrialized nations; in the next century, their successors promise to do the same for the peoples of the developing countries.

The story you are about to read is guaranteed to be entertaining, thought-provoking, and occasionally myth-destroying.

Its authors are eminently qualified to write about the electrical engineering profession as well as its associated businesses and industries. John Ryder's extensive career included industrial experience with General Electric and the Bailey Meter Co. and culminated in his position as Dean of Engineering at Michigan State University, while Donald Fink's encompassed the exciting years of television growth. Both men have long and intimate knowledge of the IEEE, having been Presidents of the Institute of Radio Engineers, one of its predecessor societies, and instrumental in the merger of the IRE with the American Institute of Electrical Engineers.

Engineers and scientists responsible for technical advances have traditionally been labeled "thing-oriented" as opposed to people-oriented. They have acquired the label understandably, since they must be skilled in mathematics, materials, and the physical sciences. Yet those technologists who have most successfully applied technology to the needs of the consumer and society at large have often become enmeshed in broader interfaces with society at many different levels.

As a result, this account of the early years of electrical engineering is as much a revelation of the human side of engineering as it is a reminder of the technical accomplishments of the profession. For example, Charles Steinmetz, the famed General Electric theoretician of alternating-current circuitry, during his 31-year career with G.E. was also a professor, a critic of engineering curricula, and an outspoken champion of civic and political causes. Other noted engineers have supported technical causes having important implications for the citizenry at large, such as the development of low-cost hydroelectric or nuclear power, or, specifically, the development of the fast breeder reactor. Some have promoted the development of modern air-traffic control and safety systems, while still others actively seek ways to employ electronic systems in peace-keeping applications.

Another factor forcing the engineer to extend his concerns beyond his own workshop is that technical developments seldom occur in isolation. The engineer builds on what has gone before and, furthermore, he is frequently not the only one working on a particular project or invention — others, some unknown to him, may be progressing as well or better. Thus, Elisha Gray and Alexander Graham Bell both pursued the development of the telephone. Thomas Edison and his contemporary, Joseph Swan, competed in seeking a practical electric light. Many had an important role in the development of radio, which culminated in Guglielmo Marconi's transmission of a wireless signal from Poldhu, Cornwall, to St. John's, Newfoundland. Subsequent advances in radio were made in a continuum, with contributions by Reginald Fessenden, Ernst Alexanderson, Lee deForest, Fritz Lowenstein, and Irving Langmuir, to name but a few. Likewise, telephone pioneers included George Campbell, Michael Pupin, Oliver Heaviside, Edwin Colpitts, John Stone Stone, and Almon Strowger.

Many physicists and materials experts were instrumental in advancing the field of semiconductors and thus paving the way for the transistor. Karl Lark-Horovitz of Purdue University was notable among them. As early as 1928 a patent for field-effect device was issued to J. E. Lilienfeld, although it was not reduced to practice.

Several workers were simultaneously on the trail of the first integrated solid-state circuit, including Harwick Johnson of RCA, Sydney Darlington of Bell Labs, and Barney Oliver of Hewlett-Packard. Ultimately, the principal credit for its development went to Jack Kilby and Robert Noyce.

Similarly, the roots of radar lie with many who probably did not foresee it as an extension of their work, among them Heinrich Hertz, Arthur Kennelly, Oliver Heaviside, Edward Appleton, and M. A. F. Barnett. Later, Sir Robert Watson-Watt of the National Physical Laboratories in England and Albert Hoyt Taylor and L. C. Young of the U.S. Naval Research Laboratory secretly and independently adapted earlier notions of measuring distance by timing reflected waves to the concept of radar itself.

Pre-nineteenth-century developments are judged to be forerunners of the modern-day computer. George Leibniz in 1703 contributed to binary algebra. George Boole in 1854 laid the foundation for symbolic logic. In the nineteenth century Joseph Jacquard and Charles Babbage produced actual hardware: punched cards and a stored program, respectively. Howard Aiken, Grace Hopper, John Atanasoff, John Von Neumann, and Claude Shannon followed, with credit for the first electronic computer in the United States going to J. Presper Eckert and John W. Mauchley.

With engineers racing toward similar or sometimes identical goals, disagreements and confrontations between experts are to be expected. An interesting conflict, although hardly the first or the most serious, took place at the first technical meeting of the AIEE in 1884 when William Preece, head of the British Postal Telegraph, questioned Edwin Houston on how he could assert that electricity flowed in "one direction rather than another." Prof. Houston's response, in substance, was that the assumption of unidirectional flow, in either direction, was an aid to circuit analysis. Among the most famous controversies was the issue of ac versus dc for electric power distribution, with George Westinghouse favoring ac and Edison adamantly opposed—a dispute the authors consider in some detail in Chapter 3, with further descriptions in a reference at the end of that chapter.

The continuous and overlapping nature of the technical developments means that questions of priority and credit are commonplace, particularly in a field changing as rapidly as this one. History reveals patent litigation that ranges from the routine to that involving long and bitter disputes. The authors give several examples. Edwin Armstrong's resources were exhausted in defending his regenerative circuit patents, when the Supreme Court in 1934 decided in favor of his opponents. In a continuing contest with RCA, Armstrong was later to receive payment in cash and stocks that would make him a wealthy man for his superregenerative circuit. Still later, he was to accuse RCA of infringing his frequency-modulation patents, a battle he lost.

Not all disputes were technical. Theodore Vail, one of the founders of the AIEE and the president of American Telephone and Telegraph, left that firm in 1887 because of a disagreement on financial matters. Vail did not return to AT&T for 10 years, when he once again assumed the presidency.

The implications of certain electrical and electronics engineering developments are so wide-ranging that their applications cannot be left solely to private enterprise. Nor can all technical issues be settled by individuals or even organizations. The government must sometimes intervene; hence the creation of regulatory bodies such as the Federal Communications Commission and the Federal Energy Administration, or agencies like the Tennessee Valley Authority.

The U.S. government's role in ending the Marconi monopoly and fostering the birth of RCA is recounted in Chapter 5. RCA's charter was based on agreements for cross-licensing radio patents belonging to AT&T, G.E., and Westinghouse, with RCA as a patent-holding company.

Hydroelectric power is another case in point. In Chapter 6, the authors review the trials of the Tennessee Valley Authority, telling how, for example, the TVA was finally permitted to burn coal to supplement its hydrogeneration in response to increased customer demand for power. The decision backfired when the price of coal soared and the TVA angered its customers when, as a result, it raised its rates.

Competitive schemes for color television were thrown into the arbitration arena when it became clear that there were at least two technically feasible plans. The FCC, as arbitrator, first selected the CBS approach, which was not compatible with existing black and white receivers. The contest, its ultimate outcome, and the role played by the National Television System Committee is described in Chapter 9.

The contributions of the practitioners of any profession are not always faultless. It may well be the professional's ethic to aspire to perfection, but errors do occur, and electrical engineering is no exception. The accident at Three Mile Island in Pennsylvania in 1979 is only one example. It was then labeled by many as an "incident," but its ramifications since may confirm the authors' characterization of it as a "disaster" (Chapter 8). TMI's effects on the curtailment of nuclear plant planning and licensing have been wide-ranging and dramatic.

A classic error was one in which the predecessor to the FCC played a role, when it assigned, in 1912, the "worthless" (below 200-meter) portion of the radio-frequency spectrum to amateur operators. These frequencies were later to prove even better for long-distance transmission than the longer waves.

A less consequential and more easily rectified mistake took place when the first long-range navigation (loran) transmitter was put on line at the Ocean City, MD, Coast Guard station. Its 100-kW pulses rang the bells of the ship-to-shore telephones of Great Lakes ore carriers.

Though it is unrealistic to expect any profession to emerge full-blown under the inspiration of one or two pioneers who recognize its need, the humble origins of most professions are little remembered. Great credit must be given to those of our predecessors who helped construct the foundations of our present industries and institutions and at the same time taught themselves what no one had yet understood.

In this regard, in 1974 this writer had the good fortune to meet Arthur Reynders, long-time IEEE member, who, then about to celebrate his one-hundredth birthday, reminisced about his entry into the profession. Only a few schools offered electrical engineering degrees in those days, so Mr. Reynders studied civil engineering, graduating from the University of Tennessee in 1895. His intrigue with electricity took him back to the university to study advanced electrical subjects, and his first postcollege job was as a surveyor for the Knoxville Street Railway Co. Later he became a draftsman for Westinghouse in East Pittsburgh and worked his way successively up the ladder through the positions of Assistant Chief Draftsman, Design Engineer on circuit breakers and switches, General Engineer on insulation, and Assistant Manager of Engineering. After 38 years with Westinghouse, Mr. Reynders retired as Works Manager of three Westinghouse plants that manufactured transformers, electric motors, radios, and refrigeration units. Mr. Reynders' experience was not unusual for pioneers of the profession. Indeed, the authors underscore the equally humble beginnings of some of the giants of electrical engineering; Steinmetz's first job in the U.S. was as a draftsman for a small electrical manufacturer in Yonkers, NY.

If the founders of the AIEE could comment on a contemporary account in *Electrical World* that described them as "electricians and capitalists," they probably would not dispute the label, a likelihood given credence by their detailed backgrounds in Chapter 3. The first AIEE president, Norvin Green, a medical graduate who entered the telegraph business in 1863, becoming President of Western Union in 1878, seemed to fit the description perfectly.

Only in recent decades has the importance of documenting and analyzing the history of the electrical engineering profession been broadly recognized. Fortunately, good records exist at a few of the larger companies, if only because they were necessary for the prosecution of patents. Other sources include the records of both amateur and professional organizations and the writings and reminiscences of members of those organizations. The pace of technical developments in the early years matched the pace of life in general — it was less hectic than it is today, and engineers seemed more inclined to preserve artifacts for posterity. On the other hand, beginning in about 1950, developments in solid-state and computer-based technology moved so rapidly that a significant amount of archival information and some important artifacts evidently were discarded or lost. The professional historians of technology who have come on the scene in recent years must be credited with developing tools to help locate, organize, and analyze documents and artifacts that can begin to give society a more cohesive picture of the first century of electrical engineering.

It was with the help of these historians as well as with the knowledge gained during their own careers and through their personal researches that the authors assembled this exciting history of engineers and electrons.

Donald Christiansen
Editor, IEEE SPECTRUM

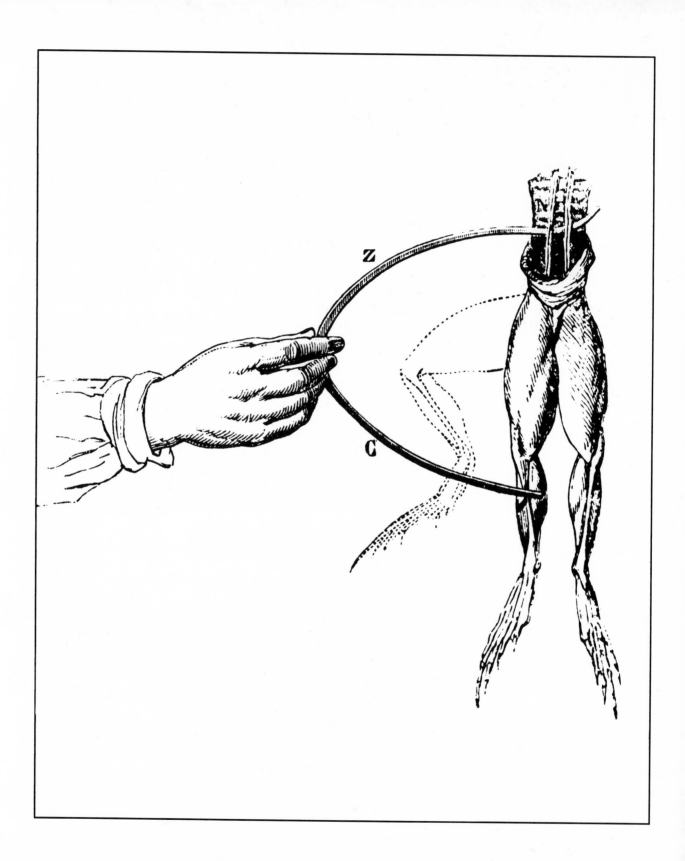

# 1

# IN THE BEGINNING

In the beginning there was the electron, unseen, unnamed, its presence recognized only by tiny forces exerted at a distance. But the mysteries created by the electron — the attractions between bodies said to be "charged," the spark to Franklin's knuckle, the apparently random effects on a compass needle — these were enough to intrigue numerous great minds of the eighteenth and nineteenth centuries. Attracted to the observation and study of the electric forces were professors, mathematicians, and abstract thinkers: men of the university and laboratory rather than men of the shop; men who laid a solid experimental, mathematical, and scientific foundation for the profession of electrical engineering. These men were more often driven by a desire for scientific prestige than for personal gain.

Franklin, as early as 1747, worked with static charges and discharges in air, noting their appearance, suggesting the existence of an "electric fluid" and reasoning that it might be made up of particles. He achieved stature in the U.S. and Europe as an experimental scientist. He was also an engineer, devising the lightning rod from his experiment with a kite in an electrical storm.

In 1785, Charles Coulomb, a French physicist, gave us the first step in scientific precision in electricity, showing that the force of attraction or repulsion between two charged spheres was inversely proportional to the square of the distance between them. The resultant mathematical statement is now known as Coulomb's law, and electrical charge is measured in units known as *coulombs*. Cavendish, in London, had discovered this same relation some years earlier, but had failed in one of the accepted duties of the scientist and had never published his work, so the credit went to Coulomb.

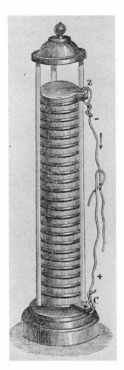

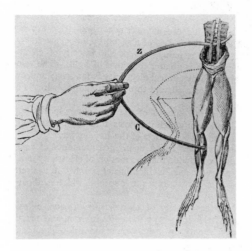

*Left: The voltaic pile.*
*Center: An illustration of*
*Galvani's discovery, taken from a*
*nineteenth-century physics textbook.*
*Z is zinc; C is copper.*
*Right: Alessandro Volta.*

Luigi Galvani, a professor at Bologna in 1786, noticed that a dissected frog's leg would twitch if touched with two dissimilar metals, copper and zinc. Galvani incorrectly assumed that electricity was produced in the frog muscle, and named it "animal electricity," to differentiate it from the "vitreous" or "resinous" electrical charges generated by rubbing glass, amber, and paper with such materials as silk and wool.

Prof. Alessandro Volta of the University of Padua disproved the reasoning of Galvani. He showed that the animal tissue was unnecessary by generating electricity with plates of two dissimilar metals separated by a salt solution. Ultimately, he used copper and zinc plates separated by cardboard soaked in a saline solution. By stacking a great number of these cells together he made the electrical "battery" in 1800. Connecting a conducting circuit to its terminals, he achieved the first continuous electric current, opening up the new field of electrodynamics for exploration. This was of such importance that Volta is honored by the use of his name for the unit of electric potential, the *volt*.

Magnetism and magnetic phenomena had been known since early times, as evidenced by such natural magnets as the lodestone and the iron compass needle. As long as only static charges were available for experiment, no relation between electricity and magnetism could be discovered, although it was sought. Only after Volta's battery made currents of electricity available could Hans Oersted, a professor at Copenhagen, find in 1820 that a current passing in a wire placed near a compass caused its needle to swing to a position at right angles to the axis of the wire. This was the first indication that electric current and magnetism were related—the basic law of electrical engineering.

Within two weeks after the news of Oersted's discovery reached Paris, Andre-Marie Ampère, a talented mathematician, observed that a coil of wire acts as a magnet when current passes through its turns and that such coils react as iron magnets do in attracting or repelling each other. He developed the equations which predict the force acting between two current-carrying conductors, and by experiment and mathematics he established the science of charges-in-motion: electrodynamics. Thus, the name *ampere* is given to our unit of electric current.

In 1827, Georg Simon Ohm, a German school teacher also working in the field of current electricity opened by Volta, gave us a simple conduction law relating potential, current, and circuit resistance, now known as Ohm's law. Again, Cavendish had experimented in this area earlier, but had failed to publish his results, so the credit goes to Ohm. The name *ohm* denotes the unit of electrical resistance.

Once it had been shown that a current produces a magnetic effect, various researchers sought a means by which magnetism might produce a current. As a major step along the way, in 1821 at the Royal Institution in London, Michael Faraday demonstrated the motion imparted to a current-carrying wire by a nearby magnet. This discovery of the force produced on an electric current by a magnetic field was the foundation for the later development of the electric motor.

At his laboratory in London, Faraday continued his search for the connection between magnetism as a source and electric current as a result. In 1831, he finally found his answer with two coils wound on an iron ring — the completion of a current to one coil deflected a galvanometer connected to the second coil. He had found that it was not the *existence,* but rather the *change* of magnetic field that was needed to create a current. He also demonstrated the principle of the homopolar generator. These experiments established the basic principles of the transformer and the rotating generator. Faraday was not a mathematician, and developed his theories by meticulous experimentation and intuition; it remained for Maxwell to formulate Faraday's results into mathematical relations some 35 years later. In recognition of Faraday's pioneering work in induction (and

*Bottom left: Michael Faraday's electrical discoveries were not his only legacy; his published account of them inspired much scientific work of the later nineteenth century. His Experimental Researches in Electricity remains today as one of the greatest accounts of scientific work ever written.*

*Bottom center: Hans C. Oersted, the first person to observe the magnetic effects of an electric current.*

*Bottom right: Faraday's rotator, in which a current-carrying wire with its end in a cup of mercury spun around an upright magnet.*

*Below: Andre-Marie Ampère siezed upon Oersted's discovery and explored it mathematically and experimentally.*

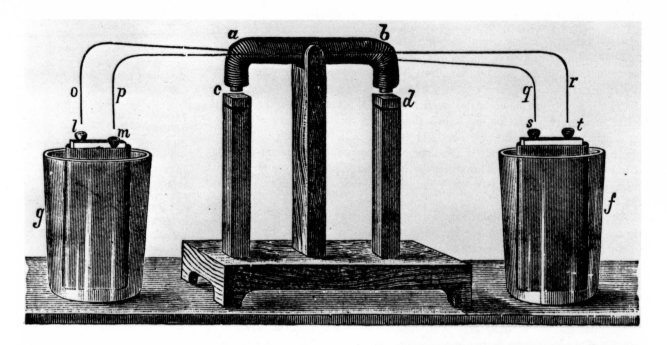

*Above: Joseph Henry's motor. As one side of the arm descended, the wires on that side made contact with the battery. The resulting current through the coils around the arm produced a magnetic field that repelled the arm from the upright bar magnet. Thus the arm would rock continuously.*

*Right: Joseph Henry in later life, when he was Secretary and Director of the Smithsonian Institution and an acknowledged leader of the American scientific community.*

electrochemistry as well), his name is carried by the unit of electrical capacitance, the *farad*.

In the United States, Joseph Henry, a teacher of mathematics and science at Albany Academy, had discovered the principle of magnetic induction, but he was unable to publish his results before Faraday. He did publish a paper in 1831 describing an electric motor. In recognition of his work the name *henry* has been given to the unit of electrical inductance.

The name of James Watt is recognized in our unit of power — the rate of conversion of energy to work — for his improvements in the efficiency of early steam engines in 1769 and later years. Watt, however, worked without the precision in heat measurement and design made possible at a later date by James Joule. Joule, a brewer of central England, was intrigued by heat measurement and sought to quantify the connection between mechanical work and heat. Self-taught, in 1840 he developed our familiar $I^2R$ relation for the conversion of electrical energy to heat in electrical circuits. His work on the

*Left: Moses Farmer, an outstanding electrical inventor of the mid-nineteenth century. Among his many achievements were motors, an early incandescent light bulb, and an important form of telegraph wire. Below: An early electric motor by a European inventor named Bourbrouze. Often, a new technology borrows heavily from existing models. In this case, the model was the steam engine, and the result was an ingenious but inefficient device.*

mechanical equivalent of heat was recognized before the Royal Society and his name is honored in our basic unit of work, the *joule*.

One more step was needed to provide the theoretical and experimental basis for the exploitation in engineering of the abilities of the electron. This was provided by James Clerk Maxwell, a graduate in mathematics from Cambridge University. In 1865, in a remarkable paper, he mathematically unified the laws of Ampère and Faraday, and expressed in a set of equations the principles of the electromagnetic field, showing the electric field and the magnetic field to be indissolubly bound together. From these universal electrical relations comes the two-arrow symbol on the badge of the IEEE, one arrow representing electric current and its associated electric field, the other the related magnetic field.

Maxwell also predicted that acceleration or oscillation of an electric charge would produce an electromagnetic field that radiated outward from the source at a constant velocity. This velocity was calculated to be about 300 000 km/s, the velocity of light. From this coincidence, Maxwell reasoned that light, too, was an electromagnetic phenomenon.

Maxwell's prediction of propagating electromagnetic fields was experimentally proven in 1887, when Heinrich Hertz, a young German physicist, discovered electric waves at a distance from an oscillating charge produced across a spark gap. By placing a loop of wire with a gap near the oscillator, he obtained a spark across the second gap whenever the first gap sparked. He showed that the electromagnetic waves behaved as light waves do; today we call them radio waves.

Faraday established the foundation for electrical engineering and the electrical industry with his discovery of induction; Maxwell, in predicting the propagation of electromagnetic fields through space, prepared the theoretical base for radio and its multitude of uses. With an understanding of the fundamental laws of electricity established, the times were ripe for the inventor, the entrepreneur, and the engineer.

Thomas Davenport of Brandon, VT, a blacksmith, used an electric motor of his own design to drive drill presses and a lathe in his shop in 1837, but general development of a motor did not follow. Adequate electric generators did not exist; batteries were the available energy source, but they were inefficient and expensive.

It was not until 1863 that Antonio Pacinotti, a young Italian professor at Pisa, made a satisfactory generator with multibar commutator. A little later, several men replaced the permanent magnet field structure with field coils. Moses G. Farmer of Salem, MA, the Siemens brothers of Berlin, and others added self-excitation. In 1870, Zénobe T. Gramme of Belgium independently duplicated and then improved upon Pacinotti's earlier designs, and in 1975 he installed generators in several French lighthouses to supply their arc lamps with current. His were the first practical and successful generator designs.

Thomas A. Edison applied knowledge of electric currents in his development of the incandescent lamp (in 1879) and its supporting power system (in the 1880's); Joseph Swan, in England, also worked toward developing a lamp. Edison's low-voltage direct-current system required heavy copper conductors; to avoid that expense, William Stanley utilized Faraday's electromagnetic induction in a transformer to provide high transmission voltages, at correspondingly reduced line current, in an alternating-current (ac) electric system at Great

Barrington, MA, in 1887. George Westinghouse of Pittsburgh, PA, with the aid of Nikola Tesla, a recent Jugoslavian emigré, further developed the transformer and ac generators into the three-phase ac generation and distribution system that powers today's world. The Tesla induction motor, also based on Faraday's principles, is the industrial workhorse of that system.

Electrical communication was on its way before midcentury. In 1837, Charles Wheatstone in England received the first patent for a commercial telegraph system; in 1844, the American Samuel F. B. Morse inaugurated commercial service between Baltimore and Washington. Another contribution was made by Morse's helper, Alfred Vail, who contrived a code of dots and dashes that minimized the time taken to send messages (see Chapter 10).

Alexander Graham Bell's telephone of 1876 was built upon the induction principle, and it took the emerging electrical profession into another new area, the practical use of time-varying currents.

The "Edison effect" (thermionic emission), discovered in 1883, remained a curiosity for two decades while the nature of electricity was further probed. G. Johnstone Stoney proposed the name "electrons" for the ultimate electric particles; J. J. Thomson, in work with cathode rays in 1897, demonstrated the existence of such particles; and Robert Millikan at the University of Chicago measured the individual electronic charge in 1911.

Founded upon the Edison effect were the two-element rectifying diode vacuum tube of Fleming (in 1904) and the three-element triode tube of Lee deForest (in 1906). Detailed understanding of the conduction process by electrons in a vacuum came from further work on the triode tube by Irving Langmuir and by H. D. Arnold in 1913.

Radio communication, based on the theories and experiments of Maxwell and Hertz, employing the tuned circuits of Oliver Lodge of England and the antennas of Guglielmo Marconi in Italy, became a field of active development. Operating at first with spark generation and broad bandwidths, R. A. Fessenden in the U.S., as early as 1906, foresaw the need for continuous waves of narrow bandwidth for voice transmission. With Fessenden's encouragement, E. F. W. Alexanderson developed high-power high-frequency rotating machines that were used for long-distance signaling by radio before high-power vacuum triodes became available after World War I. Thus, as "electrical" engineers were developing equipment to transmit power by wires over long distances, "radio" engineers were making it possible for humans to communicate through space over great distances.

An understanding of the electrical conduction process in solids was lacking in the nineteenth century because there was not yet a satisfactory model of an atom. However, knowledge was gradually being accumulated, perhaps starting with the work of Faraday in 1833, when he showed that the electrical resistance of silver sulfide was reduced as its temperature rose, an effect opposite to that observed in metals. Other work identified a group of compounds and elements — metallic sulfides, oxides, selenium, tellurium, silicon, and germanium — as having anomalous conduction properties, such as photoconductivity, negative temperature coefficients of resistance, and rectification at contacts. These materials became known as semiconductors.

In 1839, the French scientist Alexandre Becquerel had found that a voltage appeared across the junction of a semiconductor with a liquid electrolyte in the presence of light. In 1873, Willoughby Smith discovered that selenium reduced its resistance when illuminated, thus setting the stage for early television experiments. A year later, Prof. Ferdinand Braun (also a principal contributor to television) noted that rectifying action occurred when a crystal of lead sulfide (galena ore) was probed by a metal point. This discovery was the basis of one of the crystal detectors of the early days of radio, the "cat's whisker" wire that had to be so delicately adjusted to a sensitive point on the crystal before the radio signal could be heard.

Edwin Hall's 1880 doctoral dissertation (under Prof. Henry Rowland at Johns Hopkins University) and his subsequent research concerned observations that would later help to explain some of these phenomena. He reported that a voltage appeared across the opposite edges of a current-carrying metal or semiconductor immersed in a magnetic field perpendicular to the plane of the current (the "Hall effect"). The polarity of the voltage indicated that negative charges were carrying the current in some materials, but the reversed polarity showed that positive charges carried the current in other materials, particularly in semiconductors. This was years before the electron was identified and named. These discoveries were puzzles to the physicists and chemists of the day.

The strange behavior of semiconductors was not understood until the 1900's, when the quantum theory of Max Planck, the Einstein explanation of the photoelectric effect, the measurement of the electronic charge by Millikan in 1911, and the quantized nature of the electron energy levels in atoms, explained by Niels Bohr of Denmark in 1913, laid the groundwork for conduction theory and solid-state physics.

Work during World War II showed that germanium and silicon had properties suitable for development, and after the war teams of physicists, chemists, and metallurgists carried on concentrated research efforts at several laboratories, searching for possible semiconductor applications. This work culminated at Bell Laboratories in the discovery of the point-contact transistor by John Bardeen and Walter Brattain in 1947. William Shockley of the same organization worked out the theory of junctions between semiconductor materials, leading to the junction transistor.

In the 1950's, the field available to the electrical engineer widened again to include the computer and informational sciences. These areas reached back a full century to the symbolic algebra of George Boole and the early attempts at computer apparatus by Charles Babbage, both of England, and built with the more recent ideas of Atanasoff, Eckert, Mauchly, Aiken, von Neumann, and a host of others in our universities and laboratories. Claude Shannon added a further dimension by applying logic to switching circuit design and quantifying information.

Building upon the chemistry and metallurgy of the solid state, Noyce and Kilby made the most recent forward step in 1959 by integrating diodes, transistors, resistors, capacitors, and connecting leads into the body or onto the surface of a silicon chip. This advance led to the miniaturization of circuitry, a reduction in cost, and an increase in reliability for complex circuits. These integration processes were ready when the computer and information industries needed techniques that would allow expansion of the capabilities of their machines by many orders of magnitude without commensurate increases in cost.

Since the 1870–1900 era, the basis of electrical progress has been science with the relationships expressed by mathematics. New knowledge in electricity, physics, and metallurgy has preceded, supported, and now intermingles in the applications of electrical products. The electrical engineer must be able to read and understand the progress occurring across this wide spectrum of knowledge.

Today we have the electron, measured and conceived as a wave or as a particle, whose dual character is not yet understood. Nevertheless, it serves as a guiding beacon, leading today's researchers and engineers into new mysteries, new problems, and keeping the electrical profession moving into the fields of high technology of which it is a major part.

*"If I have seen further . . . it is by standing upon the shoulders of giants."*

Isaac Newton, in 1675

# 2

# DISCOVERY AND INVENTION

At the time of Faraday's discovery of induction, the speed of a ship or a man on horseback was the speed at which news traveled, little faster than Pheidippides in carrying the news of Marathon to Athens in 490 B.C. There were a few semaphore relay systems, limited to use in clear weather over line-of-sight between manned towers, but they were so expensive as to be operable only by governments. Wind-powered naval communications had limitations, too, as England had found during the American Revolution and the War of 1812. A striking illustration was the Battle of New Orleans, which occurred two weeks after the signing of the Treaty of Ghent.

At the same time, stimulated by the steam engine, which freed industry from waterfall locations, and by steam railways, which opened commerce to more than local markets, the growing industrial age was demanding fast and dependable communication.

After the discoveries of Coulomb, Ampère, Ohm, and Volta, electric current was ready for use outside the laboratory, and many saw the possibilities for meeting communication needs with the speed of an electric telegraph. The field was open to the inventors, driven largely by hopes of personal gain. Their ideas were based on a modicum of electrical theory, but they often received aid from the early scientists. Thus we see the truth of Joseph Henry's assertion that progress results from discovery followed by invention.

## DOTS AND DASHES

William Cooke, an Englishman studying anatomy in Germany, turned his thoughts to an electric telegraph after he

*Left: Charles Wheatstone, a
prominent English electrical scientist
and inventor of one of the first
telegraphs.*
*Right: The double-needle telegraph of
Cooke and Wheatstone, widely used in
England through the middle of the
nineteenth century.*
*Below: Samuel Morse as a
middle-aged man, between his art and
his invention. Appropriately enough,
the telegraph is in the foreground.*

saw some electrical experiments at Heidelberg. In London, he sought the help of Faraday, who introduced him to Charles Wheatstone of Kings College, London, who was already working on a telegraph. In 1838, they transmitted signals over 1.6 km (1 mi). Their system indicated letters by using five wires to position magnetic needles. In 1839, they extended the system to 11 km (7 mi). As a result of a discussion with Joseph Henry in England in 1837, Wheatstone later adapted electromagnets to a form of telegraph that directly indicated the letters of the alphabet.

In 1832, the young American artist Samuel F. B. Morse had his attention turned to the need for a telegraph by a fellow passenger on a return voyage from England. Morse proceeded slowly, but in 1837 he transmitted a signal for 500 m (1500 ft) over two wires. His telegraph used an electromagnet in a receiver that provided an inked record on a paper tape. In 1839, Morse was encouraged by a discussion with Henry, who backed Morse's project enthusiastically.

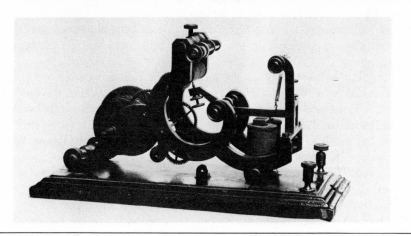

*Top: An 1844 register for receiving messages.*
*Center: The tape from the first Morse message sent on the Baltimore–Washington circuit. The words were those of Annie G. Ellsworth, the daughter of the Commissioner of Patents, who had given Morse the news that his Congressional appropriation for the line had been approved.*
*Bottom: A Morse receiving instrument of 1860.*

*The transcontinental telegraph line as it neared completion in the West. This feat put an end to one of America's most romantic episodes, the Pony Express, which had started only the year before.*

Finally, Morse sought and received a government grant of $30 000 in 1843 for a 60 km (38 mi) line from Baltimore to Washington along the Baltimore and Ohio railroad. On May 24, 1844, this line was opened with the famous message "What hath God wrought?" and Morse's system of telegraphy, using the dot and dash code and the receiving sounder devised by his assistant, Alfred Vail, was established. Since the circuit and the apparatus were simple and inexpensive, telegraph lines were rapidly built, and by 1861 the continent was spanned.

Morse soon laid a cable under the New York harbor, and a cable was laid from Dover to Calais in 1851. A cable from Ireland to Newfoundland was attempted by Cyrus Field and English associates in 1858. The mechanical design and insulation of this 3200-km (2000-mi) line were faulty, and it failed after a few weeks, but not before it carried a congratulatory message from Queen Victoria to President Buchanan on his inauguration.

William Thomson (later Lord Kelvin) had developed the theory of cable transmission, including the weakening and the delay of the signal to be expected. But it took seven years of study (from 1858 to 1865) of cable structure, cable laying procedures, insulating materials, and receiving instruments before

Field and the Atlantic Telegraph Company were able to lay a fully operational transoceanic cable in 1866. They also succeeded in dredging up the broken end of a cable lost in 1865, and completing it to its terminus in Newfoundland. Thomson was largely responsible for the research effort underlying this work.

There followed a fury of cable laying. Approximately 150 000 km (90 000 mi) of cable were in use by 1870, linking all continents and all major islands. This worldwide extension of the telegraph system made possible government at a distance, the Victorian age of the British Empire being supported on a network of telegraph wires and submarine cables.

Throughout the early history of telegraphy, Joseph Henry was an active participant. The position he took was that of the scientist in the chain of discovery and invention. He believed that once the science was ready, any knowledgeable investigator of electricity would foresee the possibility of an electromagnetic telegraph. His view was clear cut: invention should follow the discovery of principles. The Atlantic cable project, in which the laying of the first cable preceded an understanding of its basic principles, is an example of an expensive failure to act on Henry's dictum.

The telegraph business expanded as industry grew after the Civil War. But during the depression of 1873, an economic problem came to sight: simultaneous messages required individual lines, so more and more wires were needed over each route to handle the increased number of messages. By 1876, there were 400 000 km (250 000 mi) of lines on 178 000 km (110 000 mi) of

*William Thomson's work in the Atlantic cable was only one of the many achievements that made him Britain's most respected scientist during the last quarter of the century.*

*The main operating room of the Western Union telegraph office in New York in the late 1880's. The pipes are pneumatic tubes connecting this office with the smaller stations in the city, and the wires are a mechanical system for moving messages to and from the 600 operators in the room.*

routes in the U.S. This multiplicity of lines not only tied up capital, but it also darkened the skies above the city streets. Great rewards were envisioned for an inventor who could make one line carry several messages simultaneously. In fact, in 1872 Western Union paid very well for the duplex patent of Joseph Stearns, which doubled the capacity of a telegraph line. Today a single "multiplex" circuit can carry thousands of messages at the same time.

## TO SPEAK AND HEAR

The telephone was invented by men seeking the rewards offered for a multiplex telegraph system. It was in this direction that two inventors, Alexander Graham Bell and Elisha Gray, were headed in 1874. The telephone was not an invention sought after by many. In fact, the two men in the race were not even sure of their objectives until the hectic rush down the homestretch to the Patent Office. There seemed no pressing need for person-to-person communication, if only because the recipient had to be available in person to receive the message. Of the two inventors, Alexander Graham Bell was interested in speech as a science and he hoped that a telephone might lead to fortune. Elisha Gray was convinced that transmission of speech was of scientific interest only, of no commercial importance. These views arose from the different backgrounds of these men, one scientific, the other practical.

Alexander Graham Bell was the son of a professor of elocution at the University of Edinburgh. The 24-year-old Bell, after university work at Edinburgh and London, came to Boston in 1871 to teach at the School for Deaf Mutes. In 1872 he read a newspaper account of the Stearns duplex telegraph and saw an opportunity for quick fortune. By extending Helmholtz's work with vibrating reeds to generate frequencies, each frequency could be made to carry an individual telegraph message. He experimented and discussed his project with members of the Boston scientific community in 1873 and 1874.

Elisha Gray was the superintendent of the Western Electrical Manufacturing Company in Chicago (now the Western Electric Company). Early in 1874, he discovered that an induction coil could produce musical tones from currents created by a rubbing contact on a wire. As had Bell, Gray saw the possibilities for a multiplex telegraph using different frequencies, and he resigned his posi-

*A street scene in downtown Cincinnati in the 1890's. Telegraph poles carried as many as 25 cross arms, with up to ten wires on each arm. The chaos that followed heavy snow or freezing rain is hard to imagine.*

tion to become a full-time inventor. He used magnets connected to a diaphragm and located near an induction coil as receivers and transmitters.

Gray demonstrated some of his equipment to Joseph Henry in Washington in June 1874. Henry, in a letter to John Tyndall at the Royal Institution in London, expressed his opinion that Gray was "one of the most ingenious practical electricians in the country."

Bell likewise had a meeting with Henry in March 1875, in which Bell impressed Henry with his understanding of acoustical science. Henry told Bell that Gray "was by no means a scientific man," as reported by Bell in a letter to his parents. Henry encouraged Bell to complete his work; when Bell protested that he did not have sufficient electrical knowledge, Henry firmly told him to "get it."

The multiplex telegraph ideas of Bell and Gray were essentially parallel, although by 1875 Gray's equipment was further developed. However, Bell, with his scientific background, was willing to attempt the transmission of speech. Gray, the practical man, saw the rewards of multiplex telegraphy. To him, the telephone was only of scientific interest.

At that time neither man had actually transmitted speech. Although doubting telephony as a business, Gray protected his flanks by writing a description of a telephone transmitter and receiver. On February 14, 1876, his lawyer filed a "caveat" in the Patent Office. (A caveat was a priority declaration, not an application for a patent; it is no longer used.) A few hours earlier on the same day, Bell's lawyer had filed a patent application on Bell's telephone. This was done at the urging of Gardiner Hubbard, Bell's financial backer, who was aroused by rumors of Gray's progress.

Gray decided to drop his caveat in hopes of defeating Bell in the multiplex telegraph arena, so Bell was given the telephone priority. On March 10, 1876,

*Left: Alexander Graham Bell, the young teacher of the deaf whose invention made the world considerably smaller for millions of people.*
*Right: The telephone in its earliest form. In 1878, the American Physicist T. C. Mendenhall set up a pair of these in Japan, where the local royalty was delighted that the machine could speak Japanese.*
*Opposite page: An early advertisement for the telephone.*

# The Telephone.

THE proprietors of the Telephone, the invention of Alexander Graham Bell, for which patents have been issued by the United States and Great Britain, are now prepared to furnish Telephones for the transmission of articulate speech through instruments not more than twenty miles apart. Conversation can be easily carried on after slight practice and with the occasional repetition of a word or sentence. On first listening to the Telephone, though the sound is perfectly audible, the articulation seems to be indistinct; but after a few trials the ear becomes accustomed to the peculiar sound and finds little difficulty in understanding the words.

The Telephone should be set in a quiet place, where there is no noise which would interrupt ordinary conversation.

The advantages of the Telephone over the Telegraph for local business are

1st. That no skilled operator is required, but direct communication may be had by speech without the intervention of a third person.

2d. That the communication is much more rapid, the average number of words transmitted a minute by Morse Sounder being from fifteen to twenty, by Telephone from one to two hundred.

3d. That no expense is required either for its operation, maintenance, or repair. It needs no battery, and has no complicated machinery. It is unsurpassed for economy and simplicity.

The Terms for leasing two Telephones for social purposes connecting a dwelling-house with any other building will be $20 a year, for business purposes $40 a year, payable semiannually in advance, with the cost of expressage from Boston, New York, Cincinnati, Chicago, St. Louis, or San Francisco. The instruments will be kept in good working order by the lessors, free of expense, except from injuries resulting from great carelessness.

Several Telephones can be placed on the same line at an additional rental of $10 for each instrument; but the use of more than two on the same line where privacy is required is not advised. Any person within ordinary hearing distance can hear the voice calling through the Telephone. If a louder call is required one can be furnished for $5.

Telegraph lines will be constructed by the proprietors if desired. The price will vary from $100 to $150 a mile; any good mechanic can construct a line; No. 9 wire costs 8½ cents a pound, 320 pounds to the mile; 34 insulators at 25 cents each; the price of poles and setting varies in every locality; stringing wire $5 per mile; sundries $10 per mile.

Parties leasing the Telephones incur no expense beyond the annual rental and the repair of the line wire. On the following pages are extracts from the Press and other sources relating to the Telephone.

GARDINER G. HUBBARD.

CAMBRIDGE, MASS., May, 1877.

For further information and orders address

THOS. A. WATSON, 109 COURT ST., BOSTON.

at his laboratory in Boston, Bell transmitted the human voice with the famous sentences "Mr. Watson, come here. I want you," addressed to his assistant.

Bell developed instruments for exhibition at the Centennial Exhibition held that summer in Philadelphia, and these attracted great interest. Commercial installations were made in early 1877, and the first switchboard was placed in operation in New Haven, CT, in January 1878, with 22 subscriber lines. The subsequent growth of the telephone system showed that Bell had been right. People do want, and need, to talk to other people; so the telephone spread rapidly throughout the world.

## THE GENIE'S LAMP

In the first years of the nineteenth century, the renowned English chemist Humphry Davy used a large battery to produce the first electric light, an arc between two carbon rods. The high cost of the batteries made such arc lighting impractical. But in 1870, Zénobe Gramme designed a steady-current electric generator. Charles F. Brush installed street lighting with arcs around the public square in Cleveland in 1879, and in New York in 1880. This light was too brilliant and too cumbersome for home or business use, and many men tried to "subdivide the light," that is, to produce a less brilliant light source.

Thomas Alva Edison, one of those men who sought a new form of electric light, was a self-taught, highly successful inventor who had established a laboratory at Menlo Park, NJ. He had produced a carbon microphone for the telephone system, and was always interested in projects in which he could see commercial prospects. Edison was attracted to the possibilities of an incandescent lamp after a solar eclipse expedition to Wyoming in the summer of 1878. He went with George Barker, Professor of Physics at the University of Pennsylvania, who was fascinated by the work of Joseph Swan on incandescent lamps. He passed this enthusiasm to Edison, and in October 1878, Edison embarked on what was to be a hectic year of incandescent lamp development.

Edison's first lamps used a platinum filament and operated in series at 10 V each. He soon realized that parallel connections, with each lamp separately controlled, would be more practical. This increased the current to the sum of currents taken, and this made larger conductors necessary.

The laboratory staff increased as more men came in to speed the work. Among them was Francis Upton, a recent M.S. in physics from Princeton, who was added at the urging of Grosvenor Lowrey, Edison's business and financial adviser. In particular, Upton acted for Edison on mathematical problems and analyses of equipment. Edison nicknamed him "Culture."

By the fall of 1878, Edison and Upton realized that without a high electrical resistance in the lamp filament, the current in a large parallel system would require conductors so large that the investment in copper would be prohibitive. They considered the 16-candlepower gas light their competition, and they found that an equivalent output required about 100 W. Using a 100 V distribution voltage (later 220 V in a three-wire system), each lamp would need to have a resistance of about 100 $\Omega$. Although they had a platinum filament of sufficient resistance, Upton's calculations indicated that the cost of platinum would make the lamp noncompetitive.

Reading of Swan's experiments with carbon filaments in England, Edison saw the higher resistance filament that he needed. After trials of many carbonized materials, he had success on October 21, 1879, with a filament made from a carbonized cotton thread. In November, a longer-lived filament was made from carbonized bristol board, and such lamps gave about 16 candlepower with an average life of 300 h. A little later, longer-lived lamps were made with carbonized split bamboo.

*Edison and his gang at Menlo Park. Edison is in the center; to his right is Francis "Culture" Upton.*

The invention was announced by the press on December 21, 1879. Later, a public demonstration with 500 lamps was arranged at Menlo Park, and the Pennsylvania railroad ran special trains from New York for the event.

Lamps also came on board ship. The ill-fated Arctic exploration ship *Jeannette* and the *Columbia,* on the San Francisco-Portland run, were equipped with lamps. These systems operated satisfactorily, the latter serving until 1895.

# ELECTRIC LIGHT FOR
# THE WORLD

In the fall of 1881, Edison displayed with good public effect lamps and his "Jumbo" generator of 50 kW at the International Electrical Exhibition in Paris. The lengthy field structure of the Edison machines was criticized as poor magnetic design by Edward Weston, who was also manufacturing generators at that time. However, Edison and Upton had grasped the necessity for high-power efficiency, obtainable in a generator having low internal resistance. Edison claimed an efficiency of 90 percent. This figure was scoffed at by other manufacturers, who were thinking of arc light machines, which were designed for maximum power output into a fixed load and were therefore of high internal resistance to match that load. As the 1878 Franklin Institute dynamo tests had shown, these machines reached only 38 percent maximum efficiency.

This confusion between maximum output of a generator and its maximum efficiency was cleared up in 1883 by S. P. Thompson, a British engineer and writer. The confusion was between variable-voltage designs for arc lights and constant-voltage designs for incandescent lamps.

*Left: Edward Weston, an inventor who soon turned his talent to commercial electrical measuring instruments of unusual accuracy and sturdy construction.*
*Right: An early light bulb sketched in one of Edison's laboratory notebooks.*

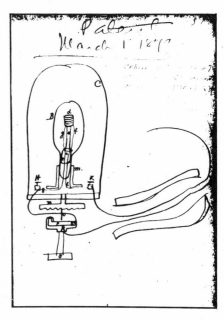

Although the use of wire-wound magnets, self-excitation, and the ring armature of Gramme were improvements, generator design was not well understood. This was particularly true of the magnetic circuit. The long magnets of Edison's "Jumbo" were an extreme example of the attempt to produce a maximum field; magnetic saturation of the structure was not understood at the time. In 1873, Henry Rowland had proposed a law of the magnetic circuit analogous to Ohm's law. But it was not until 1886 that the British engineers John and Edward Hopkinson presented a paper before the Royal Society which developed generator design from the known properties of iron. Magnetic hysteresis was discussed as early as 1881, but its full development had to wait for Steinmetz in 1892. Again, invention was waiting for discovery.

Individual steam-driven generators in stores and buildings were possible, but there were obvious problems in operating a set of boilers in every cellar, so the advantage of central generation of power for supplying lamps in mills, offices, stores, and homes was quickly seen. The first central power station was that of the California Electric Light Company in San Francisco in 1879. It was designed to supply 22 arc lamps.

The first major English central station was opened by Edison in London in January 1882, supplying 1000 lamps. The gas lighting monopoly, however, lobbied successfully for certain restrictive provisions in the British Electric Lighting Act, which delayed development of electrical illumination in the country.

Edison and Upton were also working on plans for a central power station to supply their American lamp customers. Their experimentation had shown that one horsepower of steam engine was needed to supply eight 16 candlepower lamps. This indicated a need for 1200 hp of steam power to supply 10 000 lamps. They estimated a capital investment of $150 000 for boilers, engines, generators, and distribution conductors. The buried conductors were the largest item at $57 000, of which the copper itself was estimated to require $27 000. This made clear to Edison that the copper cost for his distribution system was to be a major capital expense. To reduce the copper cost, Edison contrived first (in 1880) a main-and-feeder system and then (in 1882) a three-wire system. Annual operating expenses were placed at $46 000, and the annual income from 10 000 installed lamps was expected to be $136 000. This left a surplus for return on patent rights and interest on capital of $90 000. Financier Lowrey's boast that electric light would make Edison a very wealthy man was about to come true.

To use his system to best advantage, Edison sought a densely populated area. He selected the Wall Street section of New York, which was also close to his source of capital. He placed in service the famous Pearl Street station in New York on September 4, 1882, for which another "Jumbo" of larger size was designed. Six of these machines were driven by reciprocating steam engines supplied with steam from four boilers. They produced an output of about 700 kW to supply a potential load of 7200 lamps at 110 V. The line conductors were of heavy copper, in conduit installed underground.

The first hydroelectric station followed quickly, going into service on September 30, 1882, on the Fox River at Appleton, WI, with a 3-m (10-ft) head and two Edison generators of 12.5 kW each.

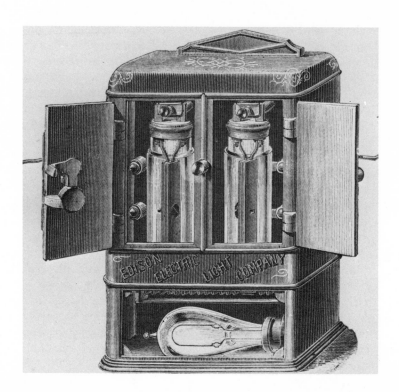

*Top: The Edison electrolytic meter.
The light bulb in the bottom kept the
solution from freezing in winter.
Right: An electric torchlight
campaign parade in New York in
1884. The lights were powered by a
steam engine and dynamo on the
carriage in the center.*

Edison had seen the necessity for the development of a complete system, and Pearl Street had conduits, switches, and fuses. To supply the electrical equipment, Edison organized five companies: the Edison Electric Light Company as the overall holding company, the Edison Electric Illuminating Company to operate Pearl Street, the Edison Machine Works for the dynamos, the Edison Electric Tube Company for the underground conductors, and the Edison Lamp Works for the lamps.

Initially, the companies billed customers for energy use by counting lamps, but Edison soon devised an electrolytic meter, a device that deposited copper from an electrolytic solution onto a plate. This plate was periodically removed and weighed, the weight of metal deposited being proportional to the current and time of use. Because the lamps' light output was not a linear function of current, there was much pressure to find some way to bill the customer for light output, not energy.

Eventually, the Edison dc system was to founder on the line cost problem. Lamp voltages and personnel safety necessarily imposed a low distribution voltage, and conductor voltage drop limited to less than 1.6 km (1 mi) the distance over which power could be economically transmitted. To serve the locations of new lamp loads, it was necessary to build other central stations in New York City. The development of the transformer ("secondary generator") and concomitant ac systems made dc distribution systems obsolete by the mid-1890's. But that obsolescence did not then—nor has it ever since—obscure Edison's central role at the dawn of the Electrical Age.

*"Perhaps the greatest discovery of Faraday's long career of scientific research was that of electromagnetic induction in the year 1831. . . . The development of electromagnetic induction has practically created electrical engineering."*
Arthur E. Kennelly, President AIEE, 1889–1890

# For Further Reading

R. V. Bruce, *Bell: Alexander Graham Bell and the Conquest of Solitude.* Boston, MA: Little, Brown, 1973.

R. Conot, *A Streak of Luck.* New York: Simon and Schuster, 1979.

D. Hounshell, "Bell and Gray: Contrasts in style, politics, and etiquette," *Proc. IEEE,* vol. 64, p. 1305, Sept. 1976.

——, "Edison and the pure science ideal in 19th-century America," *Science,* vol. 207, p. 612, Feb. 8, 1980.

T. P. Hughes, "The electrification of America: The system builders," *Tech. Cult.,* vol. 20, p. 133, 1979.

M. Josephson, *Edison.* New York: McGraw-Hill, 1959.

A. P. Molella, "The electric motor, the telegraph, and Joseph Henry's theory of technological progress," *Proc. IEEE,* vol. 64, p. 1273, Sept. 1976.

H. C. Passer, *The Electrical Manufacturers, 1875–1900.* Cambridge: Harvard Univ. Press, 1953.

G. Wise, "Swan's way: A study in style," *IEEE Spectrum,* vol. 19, p. 66, Apr. 1982.

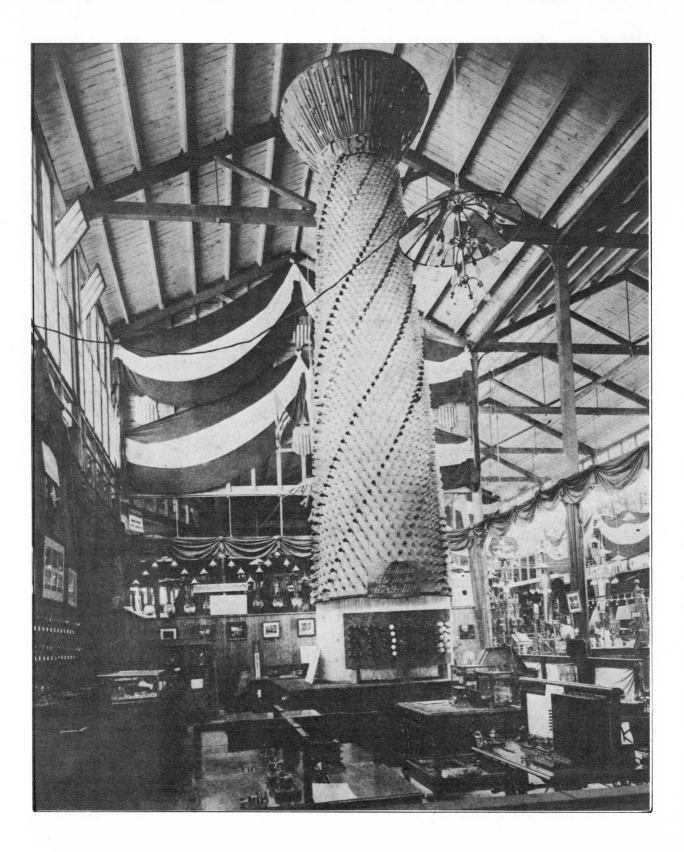

# 3

# FROM ELECTRICIANS TO ENGINEERS

## THE MAGIC YEARS

In the 1880's, the electrical industry was growing rapidly. Edison's invention and improvement of a commercial incandescent lamp, followed by his central generating station and the three-wire distribution of electric power, demonstrated the commercial possibilities of electricity. As the "Wizard of Menlo Park," Edison had secured adequate Wall Street financial backing, and had organized his companies to manufacture lamps, generators, and the elements needed for the distribution of electrical power. His success stimulated others; Edward Weston of Newark began the manufacture of motors and Charles Brush of Cleveland, while still pursuing arc lighting, began to expand in other areas. By the middle of the decade, there were over 400 private power plants, most built by Edison's company for the Edison dc system.

With the ready availability of iron and steel and of rail transport for raw materials and finished goods, the electrical industry had arrived on the world economic scene at just the right time to aid in the industrial transformation from the small isolated shop to the large integrated factory.

The electrical inventors were active in management and design for companies they had organized and often owned. Others, business men as well as those with some technical experience, had joined them. A considerable number of these early participants in the electrical business had come from the telegraph industry, usually with experience in front of a key. Largely lacking formal education in electricity, these men considered themselves "electricians."

# ORGANIZATION OF
# THE AIEE

Into this scene in the spring of 1884, N. S. Keith, an inventor and a chemist by education, introduced the concept of a national electrical society. In April he issued a circular to interested parties, which began:

"The rapidly growing art of producing and utilizing electricity has no assistance from any American national scientific society. There is no legitimate excuse for this implied absence of scientific interest, except that it be the short-sighted plea that everyone is too busy to give time to scientific, practical and social intercourse, which, in other professions, have been found so conducive to advancement."

Keith pointed out that an International Electrical Exhibition was to be held in Philadelphia in the fall of 1884, and that there should be an American electrical society present to receive foreign visitors with proper honors. He proposed to organize the American Institute of Electrical Engineers.

A paragraph later, his letter defined the area of potential membership as including . . . "electrical engineers, electricians, instructors in electricity in schools and colleges, inventors and manufacturers of electrical apparatus, officers of telegraph, telephone, electric light, burglar alarm, district messenger, electric time, and of all companies based upon electrical inventions. . . ."

The document was endorsed by a large number of those to whom it was circulated and a preliminary meeting was held in New York on April 15, at which an organizing committee was appointed. This committee included Keith and representatives from the Metropolitan Telephone and Telegraph Company, Merchants Electric Light Company, Western Union Telegraph, and the Western Electric Company.

A second meeting was called for May 13 at the offices of the American Society of Civil Engineers, 127 East 23rd Street, New York.[1] At this second meeting the proposed form of organization was adopted, as well as the name American Institute of Electrical Engineers. A slate of officers was presented by the Committee and voted upon, with Norvin Green, President of the Western Union Telegraph Company, being elected the first President. The six Vice Presidents chosen were

Prof. Alexander Graham Bell, Washington, DC
Prof. Charles A. Cross, Boston, MA
Thomas A. Edison, New York, NY
George A. Hamilton, New York, NY
Charles H. Haskins, Milwaukee, WI
Frank L. Pope, Elizabeth, NJ

---

[1] *The building is now gone and the site is occupied by a hosiery store (next door to the Gramercy Theater).*

Prominent names among the 12 Managers or Directors were Charles Brush (Cleveland, OH), Prof. Elisha Gray (Chicago, IL), Prof. Edwin J. Houston (Philadelphia, PA), Theodore N. Vail (Boston, MA), Edward Weston (Newark, NJ), and George B. Prescott (New York). Others chosen were M. L. Hellings (Key West, FL), F. W. Jones, W. H. Eckert, S. D. Field, Prof. W. P. Trowbridge (all of New York), and W. W. Smith (Indianapolis, IN).

# THE MEN OF THE AIEE

The man chosen as the first President of the AIEE, Norvin Green, age 56, was a medical graduate who had served in the Kentucky legislature and had entered the telegraph business in 1863. He became President of Western Union, the country's largest telegraph company, in 1878, and remained in that position until his death in 1893; during this period he also headed the Edison Electric Light Company for several years.

Theodore Vail had been a telegraph operator, in 1884 was with the Bell Telephone and Telegraph Company, and the following year became President of the American Telephone and Telegraph Company when it was organized. Green and Vail lacked engineering training and were listed as "capitalists."

George Prescott, a former Western Union operator, was working for Edward Weston in 1884. Frank Pope had been a telegraph operator in Massachusetts at age 17. In 1865, for Western Union, he had organized and conducted a winter survey of northern British Columbia, incident to plans for a telegraph and cable line across the Bering Strait, and then across Siberia to Moscow. This was to be an 11 000-km (7000-mi) line, but the plan was dropped after Cyrus Field's successful Atlantic cable. (Pope's journal describing the hardships of that winter on the Stikine river reads like the story of an Arctic explorer.) In 1877,

*Left: Norvin Green, first President of the AIEE.*
*Right: Frank Pope, a prominent AIEE founder. In 1869 he had formed a brief partnership with Edison to produce telegraphic inventions.*

INTERNATIONAL ELECTRICAL EXHIBITION.

FRANKLIN INSTITUTE

PHILADELPHIA, U.S.A.

*Page from the catalog of the 1884 International Electrical Exhibition.*

Pope represented Western Union as patent attorney in an unsuccessful infringement suit against Bell. Pope was to be the second President of the AIEE, in 1886–1887.

Hamilton and Jones also had Western Union experience, and Eckert was an employee of the Metropolitan Telephone and Telegraph Company. Charles Haskins was employed by Western Electric in Milwaukee.

Prof. Cross was the head of the Department of Physics at M.I.T., where an electrical engineering option had been inaugurated in 1882 under the Department of Physics. Prof. Houston taught electricity at the Central High School, Philadelphia, PA, and had joined with Prof. Elihu Thomson of the same high school to form the Thomson–Houston Electric Company, later a major competitor of the Edison Company in the manufacture of electrical machinery. Prof. Trowbridge had been a university mathematics professor, but later entered business in New York.

The early managers of the AIEE were a mixture of former telegraphers or "electricians," manufacturers, a few academics, and business men. In fact, *Electrical World* described the founding assemblage as a "group of electricians and capitalists." The background of this group had an important bearing on the nature of the discussions at the National Conference of Electricians, which was held in conjunction with the 1884 International Electrical Exhibition.

The newness of the electrical industry was evident in that seven members of the organizing group of the AIEE — Bell, Edison, Brush, Gray, Weston, Houston, and Field — were inventors or manufacturers, yet none of their companies had existed during America's Centennial Year in 1876, only 8 years earlier.

# THE INTERNATIONAL
# ELECTRICAL EXHIBITION

Sponsored by the Franklin Institute, the International Electrical Exhibition was held in Philadelphia from September 2 to October 11, 1884. The exhibits emphasized the production and application of electricity, methods of measurement, and the conductors employed, and had strong educational overtones.

The Exhibition served notice that the world had emerged into an era in which electric light and power were to affect every aspect of social and industrial activity. To the general public, it showed the heights to which the electrical inventors had moved the world. To the electrical profession, it demonstrated that the field was now too broad to be spanned by individuals like Brush, Edison, or Thomson; in the future, development would require teams of electrical engineers working within a corporate structure.

With one exception, the scientific discoveries that underlay the works of the American inventors of 1884 had come from Europe. That exception was the electromagnetic inductive effect reported by Joseph Henry. American inventions had been built upon European science by men such as Eli Whitney of cotton gin fame, Robert Fulton of the steamboat, Cyrus McCormick of the reaper, Elisha Otis of the elevator, Morse of the telegraph, Goodyear of rubber vulcanization, Singer, Edison, Elihu Thomson, and Bell. Many of these inventors had European counterparts.

Another American scientific discovery, the "Edison effect," had just been announced in 1883. Edison noted that current flowed through a vacuum in a lamp fitted with filament and a plate, and although he filed a patent on the device's use as a voltage regulator, he could see nothing else to do with it. But this discovery, nurtured by workers in science, was the seed from which would grow the electronics industry.

The Exhibition was housed in an exhibition hall of wood and glass, constructed at some financial risk by the sponsoring Franklin Institute. An old passenger depot of the Pennsylvania Railroad across the street was used as an annex. Fifteen steam boilers totaling 1800 hp provided steam for the exhibits and drove some of the prime movers. European exhibitions of those years had been housed in ornate buildings such as the Crystal Palace in London, and it was remarked by European visitors that the hall provided "exemplification of the thorough 'practicability' of Americans and of their slight regard for mere externals." Actually, the hall reflected the parsimony in government support, as Congress had only seen fit to appropriate $7500 for the conference.

Although billed as an International Exhibition, the show had only a few foreign exhibits. Exhibit materials were admitted free of duty, but foreign manufacturers had seen little sales advantage to be gained by exhibiting, since the American market was largely blocked to them by the protective tariffs of the U.S.

The exhibits were varied and reflected not only the products, but to some extent the personalities of the inventors or designers involved. Edison's exhibit was the largest and broadest in scope, and demonstrated Edison's flair for showmanship. It included not only spectacular lighting displays but a complete central station system, using a 100 kW "Jumbo" generator with the characteristically long, thin, parallel cores for the field structure. The Edison exhibit also presented the curious three-lead lamp, showing the Edison effect current through a vacuum.

The United States Electric Lighting Company showed the dynamos and lamps of Edward Weston. The motors and generators used magnetic circuits of large cross section and moderate length. These designs, and Edison's magnets, reflected the state of the art — or the state of ignorance of the magnetic art — at the time. Design efforts were dependent on individual thought, not on broadly agreed-upon principles. Weston also showed a personally made assemblage of scientific instruments, a line of equipment his company subsequently turned to.

Twenty-seven-year-old Frank Sprague was in enthusiastic attendance at his exhibit of a constant-speed self-regulating motor. His newly organized Sprague Electric Railway and Motor Company would launch its successful electric street car system in Richmond, VA, in 1887, the first truly successful line in the country. By that time there would be 21 other U.S. cities with electric traction systems.

Elihu Thomson, for the Thomson–Houston Company, displayed commercial

*Left: Grounds of the 1884 International Electrical Exhibition. Right: Poster promoting the Exhibition where the latest in electrical inventions were displayed with much showmanship.*

dynamos as well as equipment for demonstrating electrical phenomena. His lectures and demonstrations were an important educational effort and were highly popular.

The Brush Electric Company of Cleveland illustrated illumination of the home using Swan's incandescent lamps imported from England; this was a departure from Brush's previous dependence on arc lighting. It also showed that Edison's and Swan's race for preeminence in the incandescent lamp business had not been entirely one-sided.

Congress had appointed a Commission to conduct a National Conference of Electricians during the Exhibition. In planning for this Conference, emphasis had been on the scientific gains to be expected, and most of the members of the appointed Commission were teachers and scientists, notwithstanding the "Electricians" in the title of the Conference. Amid considerable friction between scientists and the more practical men, the Conference did provide for mutual discussion on electrical topics. One of these was the theory of the dynamo, but this was later compared to an "excursion into a fog" because of the limited knowledge of the magnetic circuit.

A leading physicist, Prof. Henry Rowland of Johns Hopkins University, expressed strong views calling for the support of pure science, and concluded with: "Let physical laboratories arise — let technical schools be founded — it is not telegraph operators but electrical engineers that the future demands."

E. J. Houston, as a practitioner of pure as well as applied science, tried to smooth some of the aroused feelings among the electricians still present during his concluding remarks with these words: "Theory comes as a result of experiment, and all that precedes is hypothesis."

*Left: Edison's 1884 exhibit was a display of instruments and equipment surrounding a blazing pillar of incandescent lamps.*
*Center: Frank Sprague exhibited a constant-speed self-regulating motor. It was Sprague who launched the country's first successful electric street car system in Richmond, VA, in 1887.*
*Right: Elihu Thomson was the brain behind the Thomson–Houston Company, and remained a central figure when that company merged with the Edison interests in 1892 to form General Electric.*

# THE FIRST AIEE MEETING

During the Electrical Exhibition, the AIEE held its first technical meeting on October 7–8 at the Franklin Institute.

At this two-day session, there were ten papers presented with vigorous discussion.

As it now appears, the most significant paper was the first, presented by Prof. Houston. Entitled "Notes on Phenomena in Incandescent Lamps," it described experiments undertaken with Edison's three-wire lamp of filament and plate. This was the first encounter by the audience with what we now know to be a unilateral circuit element. In that day, before electrons were thought of, those present were confused as to the direction of the current through the vacuum. By convention, based on Ben Franklin's arbitrary dictum, the current should pass from the positive plate to the negative filament. But there was no mechanism for such a transfer. William Preece, head of the British Postal Telegraph, was present. He inquired, "Professor Houston, may I ask on what grounds you assert, or anybody asserts, that electricity flows in one direction rather than another?" Houston replied, "Simply on the grounds of its being a definite idea. I don't know that anyone could prove that electricity flows more readily in one direction than another; but we have a convenient way of speaking of it as flowing from a higher to a lower potential." The matter was further confused when it was admitted that there seemed to be a much smaller current passing when the plate was negative. Today, we recognize that gas ions in the imperfect vacuum of Edison's lamp could carry this small current.

Other papers given at this first meeting covered a synchronizing scheme, underground cables and ground return, and a paper on "The Scientific City Street" by R. R. Hazard, President of the Gramme Company, generator manufacturers. This paper looked ahead to the usage of space under the streets in the not-too-distant future. Electroplating and the U.S. Patent Office were additional topics.

The organization of the AIEE had been auspiciously timed. The Institute was

*Left: Diagram from the first AIEE paper ever presented, "Houston's Notes on Phenomena in Incandescent Lamps."*
*Right: The scientific city street as envisioned by Rowland R. Hazard.*

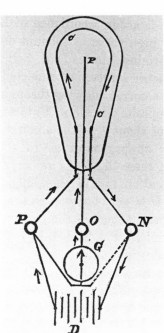

AMERICAN INSTITUTE OF ELECTRICAL ENGINEERS, SEPTEMBER, 1884.

THE SCIENTIFIC STREET

AS APPLIED TO

BROADWAY, NEW-YORK.

ready to provide a medium for the exchange of information by professional engineers, in contrast to the secrecy exercised by the inventors.

# THE ERA OF THE CENTRAL STATION BEGINS

The International Exhibition gave impetus to the construction of central electric stations near the centers of cities. At that time the load for these stations was largely incandescent lamps, operating from dusk to perhaps midnight. This represented a very uneconomical situation for these stations, since the rate per kilowatt hour had to cover the cost of idle generating equipment during the daylight hours. In England and Europe, power stations added lead-acid storage batteries on the line to help in carrying the peak loads, charging the batteries during the slack hours. The "floating" batteries on the line also improved the voltage regulation of the line and gave the service some emergency protection.

In the U.S., Edison was unalterably opposed to the use of batteries. He believed them to be a poor substitute for larger boilers and generators needed to carry the peak load. Perhaps this feeling was engendered by his unsuccessful battery experiments in 1883, but he refused to consider the improved batteries available by 1888. At any rate, Edison-controlled central stations were discouraged from battery use, but Charles Brush used them often in his arc-lighting systems, and they were extensively used in Europe.

*The induction ring of Faraday, the precursor of all transformers.*

# AC VERSUS DC

The basic problem of power distribution by low-voltage direct current was that power loss in the lines limited the service area. Electrical power is given by the product $EI$ where $E$ is the voltage and $I$ is the line current. Power transmitted at constant voltage varies as $I$, but power lost in the transmission lines varies as $I^2R$ where $R$ is the line resistance. Thus, doubling the lamp load incurs four times the line loss; that is, the line losses increase much more rapidly than the power transmitted. Since the customer voltage was fixed at 110 V for the lamp load, there was no opportunity to reduce the line losses by raising the voltage. It was necessary to reduce the line resistance $R$, but this required a major investment in copper conductors.

On the other hand, in an ac system, transformers could be used to raise the transmission voltage and reduce the line current, lowering line loss. At or near the customer's premises, using other transformers, the voltage supplied to the customer could be that required by the load of lamps or motors.

Lucien Gaulard and John D. Gibbs demonstrated an ac power system using transformers in London in 1881, and George Westinghouse of Pittsburgh acquired the American rights in 1885. There was a problem in the Gaulard and Gibbs application of the transformer to various loads. They placed the primary windings of the several transformers, feeding groups of incandescent lamps, in series across the transmission line. Hence, the voltage at the loads varied erratically as the lamps were connected to or disconnected from the transformer secondary windings. William Stanley made the first American application of alternating current at Great Barrington, MA, in 1886. He placed the primary windings in parallel across the transmission line, avoiding the voltage changes, and his installation was an immediate success. Westinghouse founded the Westinghouse Electric and Manufacturing Company and sold several ac lighting plants in that year.

Edison could have taken a position in ac power distribution, for in 1886 his company secured an option on the patents of a Hungarian transformer design. But Edison was so firmly against ac that he persuaded his companies to drop the option. It seems odd that Edison worked so hard to develop a high resistance lamp filament to reduce the line losses, and contrarily resisted so strongly the ac method by which the line losses could be reduced.

Following the initial Westinghouse successes, the Edison interests recognized that they had serious competition. Their problems were accentuated in 1887 when a French syndicate cornered the world supply of copper and the price rose from 9 cents a pound to 20 cents a pound. The copper investment and losses limited the operating radius of dc stations to about 1.6 km (1 mi). The small size of the generating plants also meant high generation costs.

In early 1888, the Edison companies went on the offensive. They emphasized the advantages of dc as the following.

1) Greater reliability, since generators could operate in parallel, a method not then worked out for ac.

2) Lack of an energy meter for ac.

3) Lack of an ac motor.

4) The suitability of dc for electroplating and battery charging.

5) The "absolute safety" of 110 and 240 V dc; the higher ac voltages would "kill a horse."

*George Westinghouse, the man who fought the battle of the currents with Edison — and won.*

The lack of an energy consumption meter meant ac stations had to be larger in generating capacity, as ac customers paid by the lamp and tended to leave lamps burning when light was not needed. However, in early 1888, O. B. Shallenberger of the Westinghouse engineering staff developed a magnetic disk meter for ac power. This was a direct-reading device, superior in operation to Edison's electrolytic dc meter. Thus was eliminated one argument of the dc proponents.

Also in early 1888, a polyphase ac motor was patented and described before the AIEE by Nikola Tesla, a recent Croatian emigré. By July 1888, Westinghouse had purchased the patents and also hired the young Tesla to help him develop his ideas. Although it took until 1892 to solve some of the problems of the induction motor and to develop the two- or three-phase system of distribution for ac power, the promise of an ultimate solution to the motor problem removed a second major argument of the Edison interests for dc. This solution followed an engineering team effort at Westinghouse, C. F. Scott and B. G. Lamme working with Tesla. The Thomson–Houston Company was not idle; there H. G. Reist and W. J. Foster contributed to the motor design solutions. These problems were ultimately solved by Lamme's squirrel-cage design.

In the summer of 1888, the Edison interests focused on the fifth argument, safety. Harold P. Brown appeared on the scene as a self-taught engineer, claiming to be independent and operating in support of public safety. (It now appears that he was aided by Edison people.) A letter from him appeared in the New York *Evening Post* asking that the Board of Electrical Control prohibit alternating voltages of more than 300 V. He described low voltage dc as perfectly safe, but said "ac can be described by no adjective less forcible than damnable." Edison personally stated that he wanted to entirely prohibit the use of alternating current or to limit alternating voltages to 300 V. He felt that the hazards associated with ac were such that the greatly increased costs of land and conductors in the dc system were worth the expense.

A bill was introduced in the Virginia legislature to ban any potential exceeding 800 V for dc, 250 V for ac. Edison appeared for the proponents of the bill at a committee hearing, but was not an effective witness because of his deafness. Opposition appeared from the state's arc-lighting companies, which used up to 3000 V dc for their series arc-lighting circuits. Since these companies would be put out of business by passage of the act, protection of local industry won the day and the bill was not favorably reported.

At Richmond and at a similar hearing in Ohio was Harold P. Brown. He demonstrated the lethal nature of ac by killing several animals before witnesses. Brown was also instrumental in having New York State adopt electrocution by ac as its method of capital punishment. The first electrocution used an alternator with poor voltage regulation, an event which seems to have been a grisly affair.

## AC IS PARAMOUNT

After the summer of 1890 the safety furor died, and Westinghouse made electrical history in that year with a 19-km (12-mi) transmission line from Willamette Falls, OR, to Portland, operating at

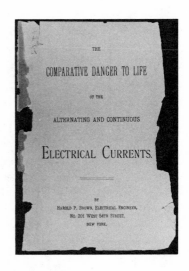

*Title page from an early book by Harold P. Brown.*

*Some of the electrical displays at the 1893 Columbian Exposition in Chicago.*

4000 V. In 1891, a 3000-V, 5-km (3-mi) line supplied a motor at Telluride, CO, at an altitude of over 3000 m (10 000 ft). Also in 1891, a 160-km (100-mi) line from Lauffen, Germany to Frankfurt was installed at 30 000 V by the Oerlikon Company and operated in three-phase fashion.

Elihu Thomson designed ac equipment that was successfully sold by the Thomson–Houston Company. The Edison General Electric Company still held viable lamp patents, but lacked access to the emerging ac equipment market. Business common sense ultimately prevailed and in 1892 the two companies merged to form the General Electric Company, which actively entered the competition for ac business. With the merger came the original buildings of the Edison Company at Schenectady. Even though he was named a director of the combined company, Edison's interest waned and he exercised no further influence in the electric power field.

Westinghouse moved to develop Tesla's polyphase system patents, that is, to produce a rotating magnetic field for the Tesla induction motor. Initially, two single-phase alternators were coupled on a common shaft, with field windings displaced at 90°. This was the system chosen to fulfill the Westinghouse contract to supply lighting power at the Chicago World's Fair in 1893, involving 12 1000-hp two-phase alternators. The use of three-phase in transmission had already been demonstrated in Switzerland by the Oerlikon Company and by the new Brown-Boveri Company. Use of three wires in three-phase transmission was advantageous because with only 50 percent more conductor area, the three-phase system would transmit 73 percent more power.

The period 1892–1893 was stimulating for electrical engineers. Polyphase systems were emerging in two- and three-phase forms and were intellectually challenging. The synchronous rotary converter had just been invented, with ac slip rings at one end and a dc commutator at the other. This machine could provide dc for street car use from ac polyphase systems. C. F. Scott of Westinghouse invented the "T" transformer connection for static conversion of two-phase to three-phase or vice versa. And, leading the way into massive ac system design was the Niagara Falls power project.

# NIAGARA FALLS — AN ENGINEERING TRIUMPH

The Niagara Falls project represented a major example of compromise in engineering design, as power plant design evolved over the years of planning.

A first agreement was necessary on the amount of water that might be diverted from the Falls without reducing their grandeur as a tourist attraction. Aesthetic considerations entered and were complicated by the lack of accessible land not devoted to park use at the Falls. This problem was resolved by choosing a site well above the falls, with generators located above deep turbine pits and diversionary tunnels returning the water to the river below the falls. Not only electricity, but also compressed air, hydraulic transmission, and wire rope proposals were considered for transmission of the generated power to markets nearby and at Buffalo.

The success of the German high-voltage line at Frankfurt altered the local thinking. Prof. George Forbes, formerly of the British Electric Light Company and a designer of alternators, was brought in as an advisor. The nonelectric and dc proposals were dropped. Forbes then started to promote his own generator design, on which bids were ultimately called for from Westinghouse and General Electric. Based partly on the firm's extensive ac experience, Westinghouse produced much of its own design and won the contract. After much debate, the specifications called for three alternators, 5000 hp each, two-phase, 2200 V, and 25 Hz. The matter of frequency was a particular problem as it affected mechanical design. Westinghouse preferred 33⅓ Hz and Forbes wanted 16⅔ Hz. These odd frequencies were the result of both a chosen water turbine speed of 250 rpm and an integral number of poles. The machines were unique in that the field structures rotated on the outside of fixed armatures, adding to the flywheel effect that helped to stabilize the hydraulic turbines.

The availability of electric power at the Falls soon attracted large electrochemical companies, such as those manufacturing aluminum and carborundum.

The first power was delivered on August 26, 1895, and the 11 000-V three-phase line to Buffalo, operating through Scott-T transformations, was energized in November.

The choice of frequency was not as arbitrary as might be assumed. It arose from the penchant of mechanical engineers to design steam engines that performed an integral number of revolutions per minute. In 1886, an available Westinghouse alternator operated at 2000 rpm and with 8 poles gave 8000 cycles per minute, or 133⅓ Hz. Transformers operated well at that frequency, and there was no flicker in incandescent lamps or arcs, but that frequency was too high when ac motors appeared at a later date. In 1889–1890 a lower frequency resulted from direct connection to steam engines, and L. B. Stilwell proposed 3600 cycles per minute or 60 Hz. This was acceptable for both lamps and motor loads.

In 1893, when the Niagara alternators were being designed, a compromise was reached at 25 Hz. This was well suited to motor applications or electrolytic refining, but was within the flicker rate of incandescent lamps. In later years it produced a very objectionable flicker in fluorescent lamps, and areas supplied with 25 Hz have now largely been changed to 60 Hz operation. With the advent of the steam turbine, the rotating speeds went to 1800 rpm for four-pole and in recent years to 3600 rpm for two-pole machines, all at 60 Hz output.

U.S. equipment is usually designed for 60 Hz, while Great Britain and most other European countries utilize 50 Hz. Much consumer equipment can operate on either frequency. This world-wide uniformity of frequency, without governmental intervention, is a victory for professional interchange and cooperation.

# THE ELECTRICIANS PASS

By 1895, when E. J. Houston retired from the presidency of the AIEE, the day of the inventors and electricians was at an end. For the next years, the professors, professional engineers, and consultants would guide the Institute. Although the initial officers and Board of Managers had been strongest in the telegraph field, a survey of the topics

*Left: The powerhouse at Niagara Falls.*
*Right: More than half of the power generated at the Falls was used by industry in the immediate vicinity.*

Table 3.1   SUMMARY OF TOPICS IN AIEE TRANSACTIONS

| | 1884–89 | 1890–94 | 1895–99 | 1900–05 |
|---|---|---|---|---|
| LIGHTING | 10 | 9 | 12 | 17 |
| CIRCUITS, DEVICES | 8 | 18 | 13 | 10 |
| TELEPHONE | 1 | 1 | 2 | 10 |
| TELEGRAPH | 5 | — | — | 6 |
| MACHINERY, TRANSMISSION | 14 | 37 | 50 | 146 |
| TRANSPORTATION | 7 | 12 | 10 | 47 |
| SCIENCE, INSTRUMENTS | 10 | 18 | 8 | 20 |
| RADIO | — | — | 2 | 3 |
| | 55 | 95 | 97 | 262 |

appearing in the AIEE TRANSACTIONS from 1884 to 1905 does not reflect that dominance (see Table 3.1). The electricians had no tradition of publication, and they represented a mature field.

The answers to the problems of the electrical industry and the technical questions facing it lay in the better understanding of electrical machines, the design of transmission facilities, and solving the problems of electrical transport. As a result of the direction taken by the technical program of the Institute, the electricians active in the founding group had largely disappeared from AIEE management by 1890.

# EDUCATION OF THE ENGINEERS

In the 1890's, a new generation began to appear on the AIEE technical programs in New York. The pages of the TRANSACTIONS frequently bear the names of A. E. Kennelly, C. P. Steinmetz, and M. I. Pupin, the latter two being graduates in mathematics and physics of European educational institutions. The electrical engineers of the 1890's were finding their professional organization, the AIEE, an efficient means of transmitting and discussing the technical knowledge being introduced by these men and by others, less prolific writers, such as Nikola Tesla. This spread of knowledge was an important part of industry's transition from rule-of-thumb at the bench to scientific accuracy at the machine. Much of the material presented at the Institute meetings in New York must have been at a mathematical level well above the abilities of the average members, but men attended the meetings to learn from the masters. This was professional education at its best.

Starting in 1892, Charles Proteus Steinmetz was one of the most prolific authors, producing over 30 major papers in a space of 30 years. Educated at Breslau and Zurich, he arrived in New York in 1889, but being crippled was almost refused entrance to the U.S. His first job was as a draftsman with Eickemeyer and Osterheld, a small electrical machinery manufacturer in Yonkers, NY. He must have been able to exceed drafting responsibilities because in January 1892, he presented to the AIEE a mathematical and experi-

mental paper on the "Law of Hysteresis." This remarkable presentation impressed Elihu Thomson and E. W. Rice, Jr. of the Thomson–Houston Company, soon to become part of the General Electric Company. Rice tried to lure Steinmetz from his employer to work for the new Company, to augment with mathematics the empirical techniques of the Edison–Thomson era. In this respect Rice was unsuccessful, but he persuaded the new General Electric President, Charles A. Coffin, that the combination of Eickemeyer's patents and Steinmetz's ability should be purchased. Thus, the General Electric Company, practically at its establishment, obtained the services of Charles Proteus Steinmetz (the Proteus was his own choice, Carl August Rudolph being the name given him by his parents).

One of Steinmetz's most far-reaching efforts occurred in 1893, when he discussed an aspect of a fundamental AIEE paper by A. E. Kennelly on impedance. Kennelly was then employed by Edison. Fearing that the use of $\sqrt{-1} = j$ in reactance terms would be lost in Kennelly's voluminous paper, Steinmetz noted that $a + jb = r(\cos \Phi + j \sin \Phi)$ showed the equivalence of the complex plane notation and the polar form in ac circuit calculations. He said, "Analysis of the complex plane is very well worked out, hence by reducing the electrical problems to the analysis of complex quantities they are brought within the scope of a known and well understood science." Since complex plane analysis did not seep into electrical engineering education for many years, his audience had to take that statement on faith. From that time, however, Steinmetz and Pupin used complex quantities in their papers, largely avoiding the cumbersome vector diagrams of other authors. Later that year, Steinmetz covered the matter more thoroughly in a masterly paper before the International Electrical Congress in Chicago.

# STANDARDS

The scientists and early experimenters corresponded across national boundaries, and recognized that electricity was international in scope. Later, it was realized that for commercial use a kilowatt must be a kilowatt everywhere, and that a motor must perform the same on both sides of national boundaries. The defining and naming of units was undertaken by an International Electrical Congress, representing major governments which could enact the accepted standards into local law.

An 1890 paper by Kennelly introduced the first proposal for the Institute to adopt a formal name — that of "henry" for the unit of inductance. This continued the practice of the profession in naming units for its famous founders. Kennelly proposed $H$ as the symbol, with lower case $h$ for millihenry, and remarked that microhenry was used so seldom as not to need a symbol! Competing unit names were "millions of centimeters," and "quadrant," the latter derived from $1 \times 10^9$ cm (the distance from the equator to the pole).

In 1891, the Institute appointed a Committee on Units and Standards with Kennelly as chairman to pursue standards efforts, especially as in 1893 the International Electrical Congress was to meet in Chicago.

Prior to this first Institute activity on units, the ampere had been defined in terms of a weight of silver deposited electrolytically by a current, the ohm as the resistance of a column of mercury, and the volt by Ohm's law $V = IR$. The entrance of ac into the field and the increasing use of mathematics in Institute papers — such as an 1892 paper that undertook the several solutions of RLC circuits, using the mathematical operators $j$ and $d/dt$ — brought out the need for agreement on names of circuit elements.

The Steinmetz paper of 1892 on hysteresis used "magnetic resistance," but Steinmetz proposed "reluctance" as its proper name. That the membership was not entirely happy with such esoteric proposals is shown by a comment from one attendee: "We have not much time in America to tackle such a subject — we see it very often in Europe . . . ."

In his 1893 paper, Kennelly proposed "impedance" for what had been called "apparent resistance," and Steinmetz suggested "reactance" to replace "inductance-speed" and "capacitance-speed" ($2\pi fL$ or $1/(2\pi fC)$) or "wattless resistance."

The year 1893 also saw the first activity of the Standards Committee in the presentation of a table of copper wire characteristics. An 1898 committee, chaired by Francis Crocker, described its own activities for machinery standardization by stating, "It aimed to define and state in as simple language as practicable, the characteristics, behavior, rating, and methods of testing of electrical machinery and apparatus, particularly with a view to setting up acceptance test standards for the electrical industry." This has remained the policy for Institute standards.

Electrical engineering had close ties to physics and science, and the first units were defined under the centimeter-gram-second (CGS) system of measurement. In 1940, the International Congress changed to a meter-kilogram-second-Ampere (MKSA) system of units. The easy interchangeability of the two bases required no fundamental change in our electrical units. The MKSA system was not generally adopted until 1946 because of World War II, but it is in common use by all today's electrical engineers.

# AN INSTITUTE BADGE

As early as 1890, some members thought an Institute badge was needed for identity. Instructions were finally given to a committee, and in 1892 it produced a design that included the Franklin kite as a basic shape, a Wheatstone bridge around the border, a galvanometer representing magnetism and induction, and a blued steel compass needle with a small superimposed amber disk. The letters AIEE were to be above the galvanometer and Ohm's law ($C = E/R$) placed below. Obviously, such a design could only come from a committee. The resultant badge had a short life, and in 1897 was replaced with the Institute badge that survived to 1963. This had a modified kite form, the letters AIEE, and two linked circles as evidence that "electricity surrounds magnetism and magnetism surrounds electricity."

# THE MEMBERS STIR

It was the custom to transact Institute business at an annual meeting, held in May. Several items from the 1893 meeting are worthy of mention. Members not living in New York City and not able to attend Institute technical meetings asked for consideration of their needs. In particular, a group of members residing in the Chicago area wanted permission to hold meetings under the AIEE banner. Financial needs were discussed, and the sum of $10 per meeting was suggested. From this discussion came the practice of same-evening meetings in New York and Chicago, the same topic being discussed at both sites with publication of the discussion of both sets of papers. Some years later the Chicago and other sections were formally organized.

It was also the annual practice for the Board to request nominations for Institute officers in advance, and to present a slate of nominees for election at this annual meeting; this slate was supposedly selected from the nominations received. The official slate in 1893 proposed T. D. Lockwood for President. A second slate was presented by petition with E. J. Houston as the nominee for President. After the balloting among the members in attendance, it was announced that Houston and the petition slate had won. In that year, the Institute membership was 717.

The AIEE was the fourth engineering organization formed in the U.S., being preceded by the American Society of Civil Engineers (ASCE) established in 1852, the American Institute of Mining Engineers (AIME) in 1871, and the American Society of Mechanical Engineers (ASME) in 1880. In England, the British Institution of Civil Engineers had been organized in 1818, and the Institution of Electrical Engineers (IEE) traced its roots to the Society of Telegraph Engineers of 1871. In 1884, there were organizations of civil engineers in Holland, Belgium, Germany, and France. Today there are organizations of electrical engineers in most European countries and elsewhere around the world.

It was proposed at the organization of the ASCE that it include civil and mechanical engineers and architects since "the line between them cannot be drawn with precision." However, the high membership qualifications and professional practice requirements established by the new ASCE precluded such a union; New Yorkers controlled the Society. This set the stage for the AIME, organized in Wilkes-Barre, PA, in 1871, with very open membership policies. When the AIEE was organized its membership requirements were more open than the ASCE's, in order to recognize the inventor–businessman, but the Board of Directors limited the election to Member grade to professional engineers for some years.

## For Further Reading

R. Belfield, "The Niagara system: The evolution of an electric power complex at Niagara Falls, 1883–1896," *Proc. IEEE,* vol. 64, p. 1344, Sept. 1976.

T. Bernstein, "A grand success," *IEEE Spectrum,* vol. 10, p. 54, Feb. 1973.

J. M. Gibson, "The International Electrical Exhibition of 1884: A landmark for the electrical engineer," *IEEE Trans. Educ.,* vol. E-23, p. 169, Aug. 1980.

*IEEE Spectrum,* "Charles Proteus Steinmetz," vol. 2, p. 82, 1965.

R. R. Kline, "Professionalism and the corporate engineer: Charles P. Steinmetz and the American Institute of Electrical Engineers," *IEEE Trans. Educ.,* vol. E-23, p. 144, Aug. 1980.

A. M. McMahon, "Corporate technology: The social origins of the American Institute of Electrical Engineers," *Proc. IEEE,* vol. 64, p. 1383, Sept. 1976.

T. S. Reynolds and T. Bernstein, "The damnable alternating current," *Proc. IEEE,* vol. 64, p. 1339, Sept. 1976.

R. H. Schallenberg, "The anomalous storage battery: An American lag in early electrical engineering," *Tech. Cult.,* vol. 22, p. 725, 1981.

——, *Bottled Energy: Electrical Engineering and the Evolution of Chemical Energy Storage.* Philadelphia, PA: American Philosophical Society, 1982.

C. Susskind, "American contributions to electronics: Coming of age and some more," *Proc. IEEE,* vol. 64, p. 1300, Sept. 1976.

D. O. Woodbury, *Beloved Scientist: Elihu Thomson.* New York: McGraw-Hill, 1944.

# 4

# MAXWELL'S PROPHECY FULFILLED

In 1865, Maxwell mathematically predicted the propagation of electromagnetic waves through space; in 1887, Hertz showed that such waves existed. Discovery having occurred, the base was laid for invention.

## MARCONI TAKES THE STAGE

On December 12, 1901, on a barren summit above the harbor of St. John's, Newfoundland, Guglielmo Marconi received a wireless signal sent from Poldhu, Cornwall, 2800 km (1800 mi) across the Atlantic. The critics had said "impossible" because of the line-of-sight nature of radio waves, but the waves did curve around the earth, refracted by the ionized layer in the upper atmosphere, a phenomenon yet to be explained by Arthur Kennelly of New York and Oliver Heaviside of England.

Marconi started to work with wireless in Bologna, Italy, but went to England with his apparatus in 1896, believing commercial opportunity to be better there. Increasing the distance over which he could receive a signal was his constant objective. He connected an antenna to the spark oscillator and this gave him greater range. He improved the Lodge coherer detector so that he could receive weaker signals. After reaching England he sent a signal over a 15 km (10 mi) water path. Then in 1899, using Marconi's equipment, crude as it was, the British navy communicated at distances up to 130 km (80 mi) over water. In the same years, Prof. Alexander Popov in St. Petersburg was also achieving some success with wireless transmission by a similar apparatus, but he lacked Marconi's drive toward commercialization of the invention.

The only tuning in the Marconi transmitter was provided by the antenna, which had a broad frequency response. Spark signals excited the antenna with pulses, which were quickly damped by the circuit resistance so the signals covered a broad frequency band. Oliver Lodge of England added tuned coupling, lowering the resistance of the antenna circuit and reducing the damping of the pulses. The more sustained oscillations reduced the frequency spread of the radiated signal. Lodge patented the use of tuning for signal selection in 1897, but he did not enforce his rights for several years and a muddled patent situation developed. Ultimately, his patent was purchased by the Marconi Company in an attempt to squeeze out rivals in supplying equipment to the ship-to-shore service.

# THE ELECTRON IS
# IDENTIFIED

In 1897, J. J. Thomson identified the electron as a negatively charged particle in a cathode-ray tube beam. The name for the particle had been proposed in 1891 by G. Johnstone Stoney. The finding of the electron cleared up the puzzle of the Edison effect, and in 1904, J. A. Fleming of England utilized Edison's effect in a rectifier to detect radio waves. The output of his diode, or two-element tube, produced an audible signal from the radio signal. The diode was stable but not very sensitive as a detector, and carborundum, silicon, and galena crystals continued to be used.

*Above: The Fleming diode as it first appeared.*
*Left: Maxwell in 1855 as a student at Cambridge. He is holding the color wheel he devised for experiments in color vision, work for which he was awarded the Rumford Medal of the Royal Society in 1860.*
*Right: Maxwell the mature scientist.*

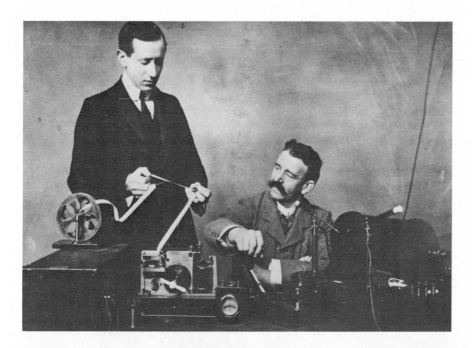

*Left: Marconi (standing) examines the recording tape from a wireless set. Below: The Brant Rock installation which housed an alternator designed by Reginald Fessenden and E. F. W. Alexanderson. They transmitted a music and voice program from Brant Rock on Christmas Eve, 1906.*

*Above: Reginald Fessenden (seated)
and his gang at Brant Rock, MA,
around 1905.*
*Right: Alexanderson with one of
his alternators.*

**Electrons & Engineers**

*Left: A 1906 field wireless set of the U.S. Signal Corps. Including the mast for the antenna, it weighed 320 pounds.*
*Below: Another Alexanderson alternator installation. This one was at Rocky Point, NY.*

Reginald Fessenden, of the National Electric Signaling Company in Boston, realized in 1901 that the spark pulses would not serve for communication by voice and that a continuous wave form of signal would be necessary. This was achievable with an arc oscillator, but more reliably so if an alternator were used. E. F. W. Alexanderson of the General Electric Company, working with Fessenden, designed an alternator of high rotational speed (20 000 rpm) giving 1 kW at 50 kHz. This was installed at Brant Rock, MA, and on Christmas Eve, 1906, they transmitted a program of music and voice. For modulation, they employed a water-cooled microphone in the antenna circuit. This transmission was received as far away as Norfolk, VA.

After a policy dispute, Fessenden left the National company, but Alexanderson continued work and scaled the machine up to 200 kW at 21.8 kHz by 1918. The machine was then to enter politics, as we shall see.

## THE DE FOREST TRIODE

In October 1906, Lee deForest presented an AIEE paper on "The Audion," a three-element tube (triode). The triode utilized a wire grid between a heated filament and a plate; a very small potential applied to the grid could control a much larger potential on the plate. It was the first electronic tube that could amplify. Far from perfect in its original form, the deForest triode nevertheless started the electronic revolution.

The original experiments had been done with tubes of imperfect vacuum, and throughout his development of the audion deForest continued to believe that a small gas content was necessary for proper operation. This limited the voltage between plate and filament to a value less than would ionize the gas. He

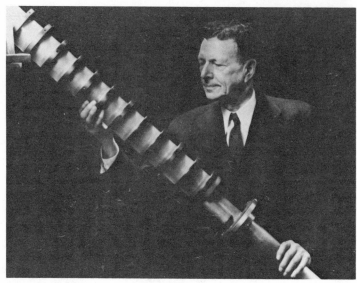

*Left: DeForest with his Audion.*
*Right: William D. Coolidge,*
*holding his multisection-type X-ray*
*tube used in a million-volt*
*X-ray machine.*

operated the tube with positive voltage on its grid, leaving the advantages of the more stable negative grid voltage to be discovered by Fritz Lowenstein in 1911, who sold his resultant patent to the Bell interests for $105 000. But the audion did amplify, and deForest set up a company to manufacture simple radio receivers, using the tube.

# INDUSTRIAL RESEARCH
# BEGINS AT
# GENERAL ELECTRIC

Just before these advances in the field of radio, a new venture in industrial management was undertaken—the research laboratory of the General Electric Company, established as a result of Steinmetz's urging at Schenectady in late 1900. Steinmetz saw competitive threats to the company's lucrative lamp business and this led him to propose a laboratory, divorced from production problems, devoted to work on scientific fundamentals underlying company products. In the fall, Willis R. Whitney, an Assistant Professor of chemistry at M.I.T., arrived as the first laboratory director. Initially, he worked only part time, but this arrangement soon became full time. In 1905, W. D. Coolidge, with a Leipzig doctorate and teaching experience in physical chemistry at M.I.T., was added to the laboratory staff.

The laboratory was then working on improvements to the incandescent lamp. Tantalum superseded carbon in lamp filaments in 1906, and Coolidge tackled the problems posed by the next step. These were associated with the use of the intractable metal tungsten. Tungsten was suited for lamp filaments because of its high melting temperature, but it resisted all efforts to shape it. Coolidge appears to have been more of an inventor than a scientist, a trait very useful in solving the problem of making tungsten ductile, which he did. Coolidge's

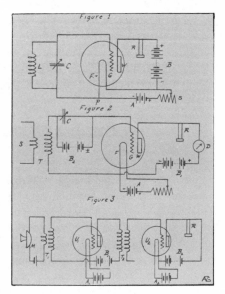

ductile tungsten filaments made a further dramatic improvement in lamp output and efficiency when introduced in 1910. Since the filaments operated at a higher temperature, they produced more light.

Another major triumph for the G.E. laboratory came from a scientific study to improve the deForest triode. This work was done in 1912–1913, after E. F. W. Alexanderson, in connection with his high-frequency alternator, had crossed trails in October 1912, with Fritz Lowenstein in the laboratory of John Hays Hammond, Jr. Alexanderson heard of Lowenstein's work with the audion and arranged for one of the tubes to be sent to Irving Langmuir, who had joined the Schenectady laboratory in 1909 from the Stevens Institute of Technology.

Langmuir believed the problems of the triode to be due to the remnant gas, which deForest felt necessary for proper operation. With a high-vacuum form of the triode, Langmuir plotted the negative region of the volt-ampere curves for the triode, found the characteristics stable, and wrote a mathematical paper explaining the tube's operation. He showed the form of these characteristics to be due to electron charge in the filament-grid space, and explained its action with his famous "three-halves-power law." Child, in England, did similar work, and we now know the result as the Langmuir–Child law of triode performance. This work turned the attention of the General Electric laboratory from lamps to vacuum tube development, and by 1918 they had produced transmitting tubes rated at 250 W output.

At the annual meeting of the AIEE in Boston in 1912, the research laboratory put its science team on show, with successive papers by Whitney, Coolidge, and Langmuir. The year 1913 brought an engineering first with a joint effort by Steinmetz of General Electric and B. G. Lamme of Westinghouse, the top engineers of their competing companies. This was a paper titled "Temperature and Electrical Insulation," which dealt with the allowable temperature rise in electrical machines.

The concept of team research that had been put forward on a small scale by Edison at Menlo Park was now well established at Schenectady and supported

*Left: B. G. Lamme, one of Westinghouse's top ac design engineers at the turn of the century.*
*Center: Some basic circuits using the Audion. (From deForest's 1913 IRE paper.)*
*Right: Irving Langmuir.*

by many accomplishments. This new management approach was valuable because it showed how the scientific advantages and freedom of the university could be combined with industrial needs and directions. Today, to further the feeling of an academic environment for the research scientist, some industrial research laboratories are built on campus-like tracts, often adjacent to major universities.

# BELL SYSTEM RESEARCH

The Bell Telephone system had gradually consolidated many small operations, but in the 1880's the engineering problems were being handled by a central Mechanical Department in Boston. This group was hard pressed to keep up with immediate needs, and it was not staffed to undertake new problems or even to understand the mathematical solutions to their transmission problems.

Confronting problems such as crosstalk, attenuation on long lines, and interference from other electrical systems required an understanding of electromagnetics. Overshadowing all was the need for understanding of the theory of transmission lines at voice frequencies. The pressure of more practical problems and lack of adequate staff prevented attack on the fundamentals, but practical advances came. Among these were the use of the common battery, the metallic circuit in preference to the single-line ground return to reduce interference, and phantom circuits creating three voice paths over two circuits. In 1893, Bell extended the transmission distance to the then-achievable limit in the Boston–Chicago line.

The basic problem, as summarized many years later by another Bell researcher, W. H. Doherty, was that "the energy generated by a telephone transmitter was infinitesimal compared with what could be generated by pounding a telegraph key. But more than this, the voice spectrum extended into the hundreds, even thousands, of cycles per second. As the early practitioners, known as electricians, struggled to coax telegraph wires to carry the voice over larger distances, they were increasingly frustrated and baffled."

The department did begin to add to staff for science studies in 1890 when they hired John Stone Stone (IRE President, 1915) from Johns Hopkins University, on Henry Rowland's recommendation. Then in 1897 George Campbell was added, who had a master's degree from Harvard and several years of advanced study in Europe. Another scientific recruit, Edwin Colpitts, replaced Stone in 1899.

In 1900, George Campbell of Bell and Michael Pupin, Professor of Mathematical Physics at Columbia University, independently developed the theory of the inductively loaded telephone line. Pupin was given the priority on patent rights. Oliver Heaviside had been the first to point out that, after the line resistance, the line capacitance most limited long distance telephone transmission. Acting between the wires, this capacitance shunted the higher voice frequencies, causing distortion as well as attenuation of the signal. He left his results in mathematical form, and in 1887 there were few electrical engineers who understood mathematics well enough to translate Heaviside's results into practical form; nor was there need to do so at that time.

Pupin drew his understanding of the physical problem from a mathematical analogy to the vibration of a plucked string loaded with spaced weights. He thereby determined both the amount of inductance needed to compensate for the capacitance and the proper spacing for the inductors. When, in 1899, the telephone system needed inductive loading to extend its long-distance lines, Pupin's patents were purchased by the Bell system at a price Pupin reported to be "most generous."

Using loading coils properly spaced in the line, the transmission distance reached from New York to Denver by 1911. This was the practical economic limit without a "repeater" or some device for regenerating the weak received signal.

Campbell's objective was always to extend the distance limits on telephone communication, but to do that he found it necessary to explore the mathematical fundamentals. His ability to do so increased the Bell Company's appreciation of in-house research, which had been only sporadically promoted. Campbell received a doctorate from Harvard in 1901, using his loading coil solution as his dissertation. He went on to develop the theory of the wave filter in 1915. This device, based on theoretical electric circuit study, enabled the "stacking" of many telephone channels on one pair of wires by filtering out separate frequency bands for each signal.

To expand the Boston group, F. B. Jewett (AIEE President, 1922–1923), an Instructor of physics at M.I.T., was added by Bell in 1904. He had a doctorate in physics from the University of Chicago and was aware of the rapid progress occurring in the physics field. In 1906 he was made head of the Department of

*Left: F. B. Jewett, one of the principal architects of the modern Bell research laboratories. Standing next to him is J. J. Thompson, discoverer of the electron.*
*Above: Michael Idvorsky Pupin, the Serbian immigrant whose loading coil lengthened telephone lines and whose Horatio Alger autobiography, From Immigrant to Inventor, won the Pulitzer Prize.*

**Maxwell's Prophecy Fulfilled**

*The completion of the transcontinental telephone line in 1914.*

Engineering at Boston, and in 1907 he aided J. J. Carty (AIEE President, 1915–1916) in moving the Boston group of engineers to New York and consolidating them with the Western Electric Engineering Department to constitute a centralized department. This coincided with a return to the AT&T presidency of Theodore N. Vail, who was one of the AIEE founders and had served as AT&T President in 1885–1887, leaving that firm after financial disagreements.

Vail's return signaled more support for basic research. This support was badly needed, as the system was about to start building a transcontinental line with the intent of initial operation at the opening of the Panama–Pacific Exposition in San Francisco held in 1915.

In-house research, under the direction of Edwin Colpitts, was authorized in 1911, and a concerted study of the repeater problem was undertaken. John Stone Stone, then serving as a consultant, called Carty's attention to the repeater possibilities inherent in deForest's audion tube. Stone, acting as deForest's agent, arranged a demonstration of the audion for AT&T in late October, 1912. The audion did not distinguish itself, due to its propensity to go into an ionized condition whenever the voltage was raised. Frank Jewett and other AT&T people had also seen a demonstration of the negative-grid usage of the audion by Fritz Lowenstein in January, 1912, although Lowenstein did not disclose his device or the circuits used.

H. D. Arnold, recently added to the staff and one of Prof. Robert Millikan's young men from the University of Chicago, was optimistic about possible improvements to the audion. He essentially solved the problems then also under study by Langmuir at General Electric. Arnold's high-vacuum tubes, with oxide-coated filaments to reduce filament heating power, were used in October 1913, in a repeater operating on a circuit from New York to Baltimore. The transcontinental line, with three spaced repeater amplifiers, was used by Vail in July 1914, on schedule. The next January, President Woodrow Wilson and Alexander Graham Bell spoke over it from Washington in the opening ceremonies of the Panama–Pacific Exposition in San Francisco.

The Bell system initially had problems with the switching of customers' lines, solved with the so-called multiple board, which gave to an array of operators access to all the lines of the exchange. Men were initially used as operators, but soon were being replaced with women. In the words of one exchange manager, "The service is very much superior to that of boys and men. They are steadier, do not drink beer, and are always on hand."

When automatic switching equipment was invented in 1889 by Almon B. Strowger, an undertaker from Kansas City, the Bell Company reacted as a monopoly sometimes does — it discounted the innovation. It took 30 years for Bell to drop its opposition. The automatic switching equipment, manufactured by the Automatic Electric Company of Chicago, operated by means of pulses transmitted by the dial of the calling telephone. Rotary relay mechanisms moved a selector to the correct tier of contacts, thus choosing the subscriber line desired by the calling party. The first installation was at LaPorte, IN, in 1892, with an improved design installed in 1900 at New Bedford, MA, for a 10 000-line exchange. This was a completely automated telephone exchange.

After noting the challenge of the new system to the capital invested by Bell, the Bell position was covered by J. J. Carty of New York Telephone in 1906: "The automatic system possesses no practical service advantage over the manual system and it contains no element sufficient to warrant any part of the extra cost which its use involves." This was not the first time that a Bell engineer looked at his company's investment in the status quo and not at the problem of supplying operators as the telephone system grew to 9.5 million customers by 1914. In 1921, when many cities were living with the inconvenience of two competing systems, the Bell interests absorbed the independent automatic companies and began a changeover to the dial system. Improved electronic switching systems are, of course, in use today.

Bell's support of basic research in the field of communications was furthered in January 1925, by the organization of the Bell Telephone Laboratories from the Western Electric engineering department in New York, then numbering more than 3600 employees. F. B. Jewett became the first President of the Laboratories, which have through the years been a source of major contributions to the communications field. The staff was augmented with people such as R. V. L. Hartley (who contributed both the oscillator circuit and communication theory), R. A. Heising (responsible for a method of modulation), and John R. Carson (who contributed single-sideband modulation).

In 1916, Bell engineers used a large array of small triodes in parallel to transmit voice by radio from Washington, DC, to Paris and Honolulu. There was a clear need for larger and more powerful tubes to open long-distance radio-telephone communication. Although Alexanderson's high-frequency alternators were used in point-to-point radio-telegraph service for many years and their low frequencies were later found useful for communication with submarines, the future of radio lay with the vacuum tube.

An unexpected research bonus came from Karl Jansky of the Bell Labs in

*Left: Armstrong's regenerative circuit as it appeared in a wireless set.*
*Right: Edwin H. Armstrong.*

1928, when he began a study of static noises plaguing the new transatlantic radio-telephone service. He found that most noises were due to local and distant thunderstorms, but that there was also an underlying steady hiss. By 1933, he had tracked the source of the hiss to the center of the Milky Way. His results were soon confirmed by Grote Reber, a radio amateur with a backyard antenna, and after the war the new science of radio astronomy rapidly developed with giant space-scanning dish antennas. This gave vast new dimensions to the astronomer's work, adding radio frequencies to the limited visual spectrum for obtaining knowledge from space.

## ARMSTRONG'S FIRST INVENTION

About 1913, Edwin Armstrong of Columbia University had evolved the regenerative circuit, using positive feedback from the output circuit of a radio detector to the input circuit. Since this connection effectively increased the input signal, amplification was increased many times. Perhaps the fundamental of Armstrong's invention was his realization that some of the high-frequency input signal existed in the plate circuit and could be fed back, a realization beyond the common view that after detection only audio frequencies were present. Such regenerative circuits were in fact oscillators, that is, generators of high frequencies, but they were operated at a critical point just short of oscillation. The circuit, which required some skill from the operator, was used until the 1930's, providing a means of truly long-distance communication.

At about the same time, Reginald Fessenden, Alexander Meissner in Germany, and H. J. Round in England all originated circuits giving somewhat similar results. A year or so later deForest also made similar claims. A patent action was started by deForest and later taken over by AT&T, which led to a long legal battle not based on the technical facts and exhausted Armstrong's finances. Finally, the Supreme Court in 1934 decided against Armstrong. The Board of Directors of the Institute of Radio Engineers (IRE) took notice of this injustice, and publicly reaffirmed its 1918 action in awarding to Armstrong the IRE Medal of Honor for his "achievements in relation to regeneration and the generation of oscillations by vacuum tubes." Because of the patent litigation, many companies had used the regenerative circuit without awarding Armstrong his royalties. The greatest use of the circuit occurred in the 1920's when the Armstrong circuit received major acclaim.

# RADIO IN CHAOS

Except for the orderly work of Langmuir and Arnold in making the triode a useful device, much of the work on radio in the decade prior to World War I was chaotic. Radio had overstepped its science and the spark equipment was inadequate. The field was beset by patent squabbles, and the period of the early triode was a heyday for the stock promoter and the quick-money artist.

Shipboard and shore stations were installed and operated as they pleased; stories of deliberate interference, or "brick on the key" activity, were frequent. Besides the stations for commercial message transmission, there were hundreds of stations operated by experimenters and amateurs using personally built apparatus and self-assigned call signs. For the experimenter, it was a field requiring small capital investment, and rewards for new circuits appeared great. Amateur equipment could be purchased cheaply. It typically consisted of a spark coil, a vibrator, and a coil of copper strap coupled to the antenna. For reception, the amateur used a homemade "loose-coupler" tuner, a crystal detector, and headphones. The enthusiast with friends in the telephone industry might even have obtained a triode tube.

The frequency of operation was a result of Marconi's predilection in the 1901–1914 years. In attempts to build a commercially dependable transatlantic service, he consistently moved to longer waves and higher power. He believed these moves to be correct, and he was followed to the longer waves and higher power by many others. Science could not yet tell him of the difference between the propagation of a ground wave and that of a sky wave, and Marconi was not one to wait for science. Although the long waves do provide communication channels which are more independent of atmospherics and solar disturbances, there is room for relatively few stations and the channels waste power. Marconi did not reverse his views for many years, and the ultimate was perhaps reached in his 1907 station at Clifden in Ireland, built to operate at 45 kHz with 300 kW.

This lowering of the radio frequency may have been unintentional at first. The antenna was the frequency-determining element in Marconi's system, and as he built larger antennas to increase the range, he incidentally lowered the operating frequency. No one knew what frequency he was using. In the first ten years of radio telegraphy, there was no means of determining operating frequency or wavelength. Because of the broad signals, a receiver could readily find the signal from a transmitter. Indeed, it was difficult to avoid it. For instance, there are two schools of thought on the wavelength used in 1901 at St. John's; In some opinions, it was about 366 m (0.8 MHz) and in others it was in the 2000–3000 m range (150–100 kHz).

In 1912, an act of Congress placed regulation and licensing of radio stations in the Department of Commerce, and some order began to appear. The licensing operation included assigning operating frequencies, and most commercial operators preferred the 400–1000 m range, a region in which they had experience. The amateurs were assigned as a group to the band of "200 m and down" because that was an essentially worthless section of the radio frequency spectrum according to Marconi's conclusions. This was an incorrect assessment of historic magnitude!

# ARMSTRONG'S SECOND INVENTION: THE SUPERHETERODYNE

In 1918, Edwin Armstrong again appeared on the scene. While attached to the U.S. Signal Corps laboratories in Paris, he developed a receiving circuit even more important than the regenerative circuit — the superheterodyne. Fessenden, in 1901, had proposed a reception method which he called the heterodyne. This name was from the Greek heteros (external) and dynamis (force). He used a steady signal generated in the receiver, mixing this with the received signal. From this process appeared a third frequency equal to the difference between the first two frequencies. If the difference was small enough to lie in the audible range (200 Hz–10 kHz), the modulation or variation of the incoming signal could be heard in the headphones. The method was not widely accepted because the local oscillators that generated the constant signal were not sufficiently stable.

What Armstrong invented was a great improvement on the heterodyne, justifying the name superheterodyne. It, too, used a local oscillator frequency, mixed this with the incoming signal to produce a much lower (but not yet audible) intermediate frequency (IF). This intermediate frequency was then amplified at the lower frequency and great selectivity between signals could be achieved in the tuned circuits attached to these lower frequency amplifiers. The local oscillator frequency had only to be stable enough to place the intermediate signal in the frequency band passed by the amplifiers. Thus, Armstrong overcame the limitation of the heterodyne method. After much amplification in such IF amplifiers, detection followed to make the signal audible. A very considerable increase in overall amplification was possible with this receiver, and today it is the circuit universally used in almost all radio, television, and communication receivers.

This was Armstrong's second great invention, one from which he would receive some financial reward. Two others, superregeneration and the use of frequency modulation, were to follow.

Simple receivers and transmitters with crystal or diode detectors and audio amplifiers were used on the ground and in the air in World War I. Except for Armstong's superheterodyne, which was not put into production until 1924, there was little improvement in the circuits used. However, industry learned how to produce vacuum tubes in quantity, production of one type approaching one million a year. These tubes had to yield reproducible results when placed in receivers and transmitters, and standardization of construction was difficult.

## THE GREEKS HAD A WORD FOR IT

The names diode and triode are used to designate two- and three-element tubes. These words were derived from the Greek *di*, *tri*, and *odos*, the last meaning a path or way. Starting about 1913, Greek-derived trade names were applied to the electron tubes produced by the General Electric Company, following suggestions from Langmuir, Saul Dushman, and A. W. Hull (Hull was a former scholar of Greek). This era has been spoken of as "Graeco-Schenectadian," in which names such as *pliotron* (*plio* meaning more, and *tron* as a tool), *magnetron* (for a combination of a magnet and an emitting filament), *phanotron* (*phano* meaning shining), and *thyratron* (*thyra* meaning door) were used. In Pittsburgh, Westinghouse was not to be outdone. That company coupled Latin *ignis* to Greek *tron* to form *ignitron* for a mercury vapor switch tube. This name was a matter of concern to some etymologists, as was deForest's *audion*, since the words mixed Latin and Greek roots.

## MEANWHILE ON THE SOCIETY FRONT

During his term in office, 1902–1903, AIEE President C. F. Scott (then at Westinghouse) made a practice of asking for advice from AIEE members. Monthly meetings had started in Chicago, on the same date and using the same discussion topic as the New York meetings, and members elsewhere were interested in extending this technical service to other geographical areas. A Committee on Local Organization was established, and sections were authorized in cities where there were sufficient interested members. After Chicago, groups in Cincinnati, Denver, Philadelphia, Pittsburgh, and St. Louis were organized in 1903.

Members also asked President Scott for committees to follow the progress of different technical areas as specialization occurred. Again, President Scott responded, and two technical committees appeared in 1905, the Telegraph and Telephone Committee and the High Tension Transmission Committee. The former dropped from sight from 1906 to 1908, but then picked up its responsibilities. By 1915, there were 13 technical committees; by 1962, the number had reached 51.

Members also expressed interest in organizing, as part of AIEE, electrical engineering students on college campuses. The Board of Managers responded and the records show that 13 student organizations were recognized in 1903.

The fact that these changes were suggested by the membership at large, aided by a dynamic president, seems to indicate that the AIEE Board had been concerned only with activities in New York. Local sections were not authorized to hold their own programs until 1910; fear had in fact been expressed by members that the local groups would rapidly run out of discussion topics!

# THE ENGINEERING
# SOCIETIES BUILDING

At the 1902 annual dinner of the Institute, Andrew Carnegie was the speaker. President Scott broached the subject of the need for a single building to house the engineering societies, to bring about greater professional unity in engineering. The next day Carnegie invited Scott and others to call on him. He wished to see their plans for the building, which were nonexistent. The next week they went back with ideas somewhat solidified, and Carnegie suggested a donation by him of $1 000 000, "more or less." In 1904 the "more" became an additional $500 000 to complete the building.

A site was chosen at 33 West 39th Street in New York, and plans were made to accommodate the ASCE, AIME, ASME, and the AIEE in a 13-story building. These organizations were to become known as the "Founder Societies," the first societies in the engineering field in the U.S. It was to this group that Carnegie made his gift. The United Engineering Trustees were the body organized to build and operate the building.

The civil engineers (ASCE) soon asked to drop out of the project. The society at that time occupied an adequate building of its own, and many of its members

*The ceremony at the cornerstone-laying for the Engineering Societies Building. Andrew Carnegie is in the center of the group.*

*The Engineering Societies Building.*

felt that if unity was desired, it should occur within the ASCE. They viewed the ASCE, the first engineering society, as including all nonmilitary engineers. Some of the members feared that the joint action on the new building was a way of superseding the ASCE. A mail vote was taken on the question and a substantial majority voted to stay out of the venture. President Scott, by inviting several of the smaller societies to join, managed to show sufficient unity to maintain Carnegie's interest in completion of the project. The building was dedicated in April 1907, and it served until the move to the new engineering building on East 47th Street, in 1961. In 1917, the civil engineers decided to come into the building, paying $250 000 to finance two additional floors for their use.

# PROFESSIONALISM
# IN THE AIEE

Beginning in 1912, the engineering profession passed through a time of unrest. Members expressed a general dissatisfaction with the role of engineers in their societies. Actually, the problem seems to have concerned the civil engineers and the mining and metallurgical practitioners much more than it affected electrical engineers. The two viewpoints may have arisen from differences in employment practices. Most electrical engineers worked for large corporations, with only a small fraction employed as "professional" engineers or consultants. In civil engineering, the situation was reversed, so the ASCE looked upon itself as a society of truly professional engineers. The AIME allowed corporate membership, and the unrest there arose from the consulting mining engineers and metallurgists who protested against corporate control of their society.

Professionalism for nonelectrical engineers resulted from the shift from engineering as a craft to a technology based in science. The professional engineers were brought in to explain and interpret the scientific complexities. But electrical engineering was not based on a craft — it was a child of science — and an understanding of the physical basis of things, the complexities of alternating current, and the higher frequencies, was necessary for everyday work in the field. In this sense, those employed full time in engineering departments and laboratories were professionals. In electrical engineering, there has been less need for a "consultant" to interpret the scientific questions for the client; electrical engineers have had to understand electrical science or fail in their work.

The AIEE was founded and led in its early years, not by professional engineers, but by businessmen who had moved from engineering design to management. Such leadership had become an Institute tradition (see Appendix I). In 1913, when the total membership was 7654, President Ralph Mershon expressed concern about this problem when he deprecated the tendency of the Institute to award its high offices to persons whose "chief claim to fame arises from activities in fields other than that of electrical engineering as defined by the Institute's constitution."

# MEMBERSHIP
# REQUIREMENTS

Unity among the engineering societies was still urged, but differing membership requirements made unity difficult. Before 1912, AIEE membership was in two grades, Associate and Member. An individual entered as an Associate and later applied for advancement to Member after meeting the requirements of "ability to design and direct engineering work, and with responsible charge of engineering work for two years." The Board maintained control over the Member level by limiting the transfers to Member grade to about 50 per year. The membership situation in the several

*Left: Robert Marriott.*
*Right: John Stone Stone.*

*Left: Alfred Goldsmith.*
*Right: John Hogan.*

Table 4.1    1905 ENGINEERING SOCIETY MEMBERSHIPS

|  | FOUNDED | HONORARY MEMBERS | MEMBERS | ASSOCIATE MEMBERS | ASSOCIATES | JUNIOR MEMBERS | TOTAL |
|---|---|---|---|---|---|---|---|
| ASCE | 1851 | 9 | 1795 | 903 | 127 | 367 | 3201 |
| AIME | 1871 | 7 | 3483 | — | 190 | — | 3680 |
| ASME | 1880 | 19 | 1915 | — | 237 | 609 | 2780 |
| AIEE | 1884 | 2 | 481 | 2851 | — | — | 3334 |

engineering societies is illustrated in Table 4.1, for 1905, taken from an address by President J. W. Lieb (1904–1905).

The comparatively small number of Members in the AIEE shows the effect of the limitation by the Board; the AIEE was at that time a society dominated by an engineering elite.

In 1912, in an attempt to open up the Member classification, its requirements were reduced to "ability to design and take charge under general supervision." To differentiate between the new Members and the old, a Fellow classification was established, to be applied for with five certifications of the applicant's record. It was expected that those who had been fully qualified as Members under the old regulation would apply for transfer to Fellow. At the same time, the requirements for the Associate grade were modified to allow the entrance of businessmen in electrical work.

This change in definition of Member status did not greatly affect the membership divisions in the Institute. The rate of Board transfers to Member continued at only a few hundred a year, and transfers from Member to Fellow were less than 100 a year. Overall growth was small until World War I, when total membership reached 10 000.

# ORGANIZATION
# OF THE IRE

In 1912 occurred a little-noticed event that was in time to have a major effect on the electrical profession. This

*The first annual convention of the IRE was held at the Waldorf-Astoria Hotel in New York in 1926.*

was the founding of the Institute of Radio Engineers by merger of the Society of Wireless Telegraph Engineers (SWTE), started in Boston in 1907 by John Stone Stone, and the Wireless Institute, established in New York in 1908 by Robert Marriott. Because of changes in company affiliations and locations that affected internal loyalties, both societies were struggling to hold their memberships by 1912. A preliminary discussion was held by Alfred Goldsmith of City College of New York, Robert Marriott, consultant, of the Wireless Institute, and John V. L. ("Jack") Hogan of the SWTE. On May 13, in a meeting in Fayerweather Hall at Columbia University, the Institute of Radio Engineers was organized, with Marriott as first President and Fritz Lowenstein as Vice President. Jack Hogan had been deForest's assistant in some of the early triode work; Lowenstein has already appeared in these pages. In 1912, Hogan held a patent on single-control tuning of a radio receiver, whose value would become apparent in later years.

From 46 members at the founding, the membership roll had grown to 231 by January 1914. Three Members and ten Associates of the AIEE were listed among the 46 original members. Marriott states in his autobiography that he had a helpful conversation with Ralph Pope, who was Secretary of the AIEE at the time, but no official notice was taken of the formation of the IRE by the AIEE. It is probable that the IRE founders felt that they would not receive much attention if they joined with the AIEE, which was then a growing, 28-year-old organization with a preponderant interest in electrical machinery.

That this was a reasonable conclusion is shown by the publication record of the AIEE. In the 28 years from 1884 to 1912, the AIEE had published over 750 papers on electrical machinery, transmission, and electrical transportation, and only seven papers on radio subjects.

The IRE predecessor societies did not delimit their activities by having "American" in their titles, and a move to add it in the IRE was resisted. The argument was that the IRE interests would be international in scope, as was radio itself. For an emblem, a general and perhaps perpetual symbol was desired. Since radio deals with electromagnetic energy, a representation of electric and magnetic forces was chosen, displayed by a vertical arrow and a surrounding curved arrow, as in the right-hand-rule convention. The letters I, R, and E were added, leading to a triangular badge. It is interesting that both AIEE and IRE recognized this fundamental electromagnetic relation in their badges, although with differing symbolism.

The PROCEEDINGS OF THE IRE was published on a quarterly basis almost immediately upon organization; it became bimonthly in 1916 and monthly in 1927. Dr. A. N. Goldsmith served as Editor from the inception of the PROCEEDINGS until he became Editor Emeritus in 1954, continuing to his death in 1974. Following procedures borrowed from scientific journals, Goldsmith provided rapid publication of fundamental and archival papers in the PROCEEDINGS.

The first IRE Section outside of New York was organized in Washington, DC, in 1914, and a Canadian Section was established at Toronto in 1925. Outside of the U.S. and Canada, the first Section was that in Buenos Aires in 1939. Thus the foresight of the founders in recognizing the international character of radio was justified.

The IRE initially recognized two membership grades: Member for active workers in radio, and Associate for those interested in, but not professionally engaged in the field. This made room for the great number of amateur radio operators, basement experimenters, and technicians attracted to the mysteries of the radio field. It was felt that as the Associates learned and progressed, they might move up to Member grade.

The grade of Fellow was established in the IRE to recognize outstanding achievement, and after 1940 was available only upon invitation from the Board of Directors. The grade of Senior Member was later established for those with responsible professional experience.

With the advent of radio broadcasting in the 1920's, the years of rapid growth for the IRE were at hand. It is to this active arena that we now turn.

## For Further Reading

H. Aitken, *Syntony and Spark: The Origins of Radio.* New York: Wiley, 1976.

A. M. Angelini, "Marconi's achievement: A systems outlook," *IEEE Spectrum,* vol. 11, p. 47, Dec. 1974.

J. E. Brittain, "The introduction of the loading coil: George A. Campbell and Michael I. Pupin," *Tech. Cult.,* vol. 11, p. 36, 1970.

A. Douglas, "The crystal detector," *IEEE Spectrum,* vol. 18, p. 64, Apr. 1981.

J. W. Hammond, *Men and Volts.* New York: Lippincott, 1941.

*A History of Science and Engineering in the Bell System* (3 vols.), Bell Telephone Laboratories, Inc., 1975–1982.

L. Hoddeson, "The emergence of basic research in the Bell Telephone system, 1875–1915," *Tech. Cult.,* vol. 21, p. 512, 1981.

E. T. Layton, *The Revolt of the Engineers.* Cleveland, OH: Case-Western Reserve Univ. Press, 1971.

A. A. McKenzie, "The three jewels of Marconi," *IEEE Spectrum,* vol. 11, p. 46, Dec. 1974.

G. Shiers, "On the origins of electron devices," *IEEE Spectrum,* vol. 9, p. 70, Nov. 1972.

G. Wise, "A new role for professional scientists in industry: Industrial research at General Electric, 1900–1916," *Tech. Cult.,* vol. 21, p. 408, 1980.

# 5

# THE ELECTRON SINGS A GOLDEN TUNE

At the conclusion of World War I, wireless — or radio, as it was becoming known — entered a new era, that of continuous wave transmission. It was made possible with the Alexanderson alternator and colored with the rosy promise of the vacuum tube.

Spark was dead.

## THE MARCONI MONOPOLY

The postwar period also put an end to the near monopoly held by the American Marconi Company on point-to-point radio communication in the U.S. and on board ships. The Marconi position had been attained by manipulation of patent ownership and exploitation of the Lodge tuning patent, an inescapable component of station design. The practice was to sue for infringement, and after the case was won by Marconi, to follow with a bid for the depreciated assets of the defeated defendant, patents as well as stations.

The postwar point of attack of American Marconi was with the Alexanderson alternator. The Marconi transmitting station at New Brunswick, NJ, had obtained the first 50-kW Alexanderson machine just before the station was taken over by the U.S. Navy in April 1917. When a 200-kW 25-kHz alternator was completed in 1918, it was also installed there. This machine was used to transmit President Wilson's ultimatum to Germany in October 1918.

# RCA IS ORGANIZED

After the war, the U.S. Navy was slow to release its hold on radio operations, then confined to point-to-point message service. There was an argument under way in governmental circles: Should radio be a government-owned monopoly, as it was to be in many countries, or should it be reopened to private ownership and development?

The IRE joined this argument on the side of private development, reminding Congress that the radio field still faced such important technical challenges as long-range telephony, selectivity problems, and static elimination. The IRE statement pointed out that government interference would impede technical creativity in solving these and other problems.

A crisis occurred in 1919 when the British-owned American Marconi Company proposed to buy the rights to the Alexanderson alternator from the General Electric Company, intending to maintain their monopoly position in the radio field. In Washington, a government monopoly might be considered a matter of policy, but certainly not if that monopoly was to be held by a foreign government. Seeking quick action, President Wilson turned to the General Electric Company to maintain American control of the radio industry. General Electric purchased the American Marconi Company for about $3 000 000, and proceeded to organize a new entity, the Radio Corporation of America. A few months later, AT&T joined RCA by purchase of stock, giving the new corporation control of the deForest triode patents as well as those of Langmuir and Arnold.

In 1920, Armstrong sold his regenerative-circuit and superheterodyne patents to Westinghouse for $335 000, which helped him to meet the legal bills he was incurring as the deForest suit on regeneration dragged on. Armstrong retained

*Left: A pre-1920 radio enthusiast with a fairly sophisticated receiving set.*
*Right: A pioneering wireless station antenna. (The figure at the top may be Armstrong.)*

GENERAL VIEW, WIRELESS STATION
ARLINGTON VA JULY 2-12

*Antennas and buildings at the Arlington, VA, wireless station in 1912.*

a right to license his regenerative detector circuit for "amateur and experimenter" use and this later brought him some royalties.

Westinghouse, having entered the radio field late, now had a bargaining position with the Armstrong patents, and entered the RCA group in 1921. As minority holders were United Fruit and Wireless Specialty, marine users of radio. The organization of RCA helped to clarify some of the patent arguments. From that time on, government was to control radio only through licensing of stations and allotment of frequencies. Otherwise, the industry was free to develop on its own pattern.

The Radio Corporation, born with a governmental blessing, was founded on an agreement that called for cross-licensing of AT&T, G.E., and Westinghouse patents in the field of radio. While this arrangement freed the industry of its internecine patent fights, it also placed tremendous power in one corporation. Actually, patent issues were not as smoothly resolved as it might have appeared. With only small need for capital, it was easy for small entrepreneurs to enter the radio manufacturing business and infringe major patents, with the belief that technical advances in the field would make a patent of small value before a legal chase could be organized. This unforeseen threat to RCA's corporation planning was real; by 1922, there were over 300 set manufacturers paying little attention to RCA's patents. During the 1920's, there were over 1000 manufacturers in the business, although only a few survived the intense competition.

Among the three major stockholders of RCA in 1921, the licensing agreement seemed to parcel out the opportunities rather neatly. AT&T was assigned the manufacture of transmitting and telephone equipment. General Electric and

Westinghouse, based on RCA holdings of Armstrong's patents, divided the receiver manufacturing field with 60 and 40 percent, respectively. RCA, owning patents, but forbidden to have manufacturing facilities for ten years, was to operate the transatlantic radio service; it was otherwise a sales organization.

# THE SARNOFF SAGA

RCA had another major asset in David Sarnoff, its commercial manager, who had come from a similar position with American Marconi. Brought to the U.S. from Russia at the age of nine, Sarnoff started as an office boy with Marconi, learned radio operation in night school, and served as a radio operator on board ship. Sarnoff, a born entrepreneur, was ready to climb the ladder to ultimate control.

Robert Marriott has said (in his autobiography) that in 1914, Sarnoff stated to him that he was leaving engineering work to go into sales, as "engineering was where the money went out, and sales was close to where it came in."

*Right: The KDKA broadcast room. Below: In 1921, Sarnoff led this group on an inspection tour of the RCA transoceanic station at New Brunswick, NJ. The identifiable faces include Sarnoff himself (fourth from left), Albert Einstein (eighth from left), C. P. Steinmetz (center), Alfred Goldsmith (right of Steinmetz), Irving Langmuir (next to Goldsmith).*

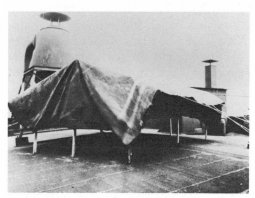

*Above: The opening of WJY-WJZ at Aeolian Hall, New York, on May 15, 1923. Alfred Goldsmith is at the microphone; David Sarnoff is seated directly before him.*
*Left: This rooftop tent sheltered some of KDKA's transmitting equipment.*

Sarnoff also was active in early IRE work, succeeding Goldsmith as its secretary in 1917.

In 1915, Sarnoff, as Assistant Traffic Manager of Marconi, had written to his General Manager with a suggestion that a receiver be designed "to bring into the home by wireless . . . lectures at home, events of national importance . . . baseball scores." Tuning was to be by a "single switch or a button." He proposed that programming costs be met from the manufacture and sale of radio receivers.

As was to be true of many of Sarnoff's ideas, the suggestion was well beyond the capabilities of the radio art at the time, and was disregarded by the RCA engineering staff. But Sarnoff became President of RCA in 1930, and after that similar proposals were made with more force and work undertaken by the RCA engineers. The most notable example was Sarnoff's "request" in 1948 for development of a color television system that would be compatible with the then existing black and white service (see Chapter 9).

Alfred Goldsmith — Professor at City College of New York, consultant to the

General Electric Company, cofounder of the IRE and Editor of the PROCEEDINGS OF THE IRE — also had transferred from the Marconi Company to RCA, as Director of Research. At first the RCA laboratory was located at City College. In 1922, some of the work of the laboratory was described as testing sample receiver designs delivered from G.E. and Westinghouse, improving the design of transmitting stations, and the development of a radically new form of "radiola" phonograph.

# BROADCASTING

By 1920, the art of the crystal detector and the simple tube receiver was in place, and an initial audience for broadcast programs was provided by the thousands of home-trained radio amateurs and set builders, each with neighbors drawn to hear music extracted from the air waves, themselves to become eager purchasers of receiving equipment.

The traditional story of radio broadcasting starts with the new station KDKA of the Westinghouse Electric and Manufacturing Company in East Pittsburgh, and with its election-night broadcast of returns of the election of Warren G. Harding as President in November 1920. This broadcast brought well-deserved publicity for KDKA and helped establish radio broadcasting. Actually, there were other and earlier efforts in transmitting speech and music for entertainment, notably WWJ of the *Detroit News* in August 1920. The art went back to Fessenden in 1901, whose broadcast was heard by many operators on ships at sea. In England, the first broadcast license was not issued until 1922, due to involved governmental controls.

The RCA agreement had not envisioned broadcasting. The agreement was intended to preserve the status quo in point-to-point message service. But that all changed when the public literally created broadcasting through its insatiable desire for home entertainment. Added to this was the new fad of long distance reception. To have heard a station across the continent insured a badge of merit when discussing the event on the job the next day. This created a heavy demand for equipment, which swamped the manufacturers and the radio stores, including departments in Kresges and Woolworth's. Local amateurs turned to building receivers; commercial set manufacturers paid no attention to RCA's patent structure, although a few paid Armstrong royalties on his regenerative circuit patent. By 1922, there were over 500 broadcasting stations, mostly assigned to operation on one frequency, but actually operating over a broad frequency band, each choosing a frequency that would avoid interference from neighboring stations. These operations centered around 360 m wavelength.

Manufacturers such as Atwater–Kent, Crosley, Grigsby–Grunow, and Chicago Radio Laboratories reaped the early rewards. Simple sets used crystal detectors, but more expensive models employed regenerative detectors and two stages of audio amplification with deForest and other makes of tubes. On local stations, the larger sets could operate simple, magnetically driven, horn loudspeakers, around which the family would gather to hear the music (and the occasional squeal of the maladjusted regenerative receiver).

In all this early ferment, RCA could only watch. It could not enter the competition under the terms of its founding. So RCA was tied to the slow-moving giants, G.E. and Westinghouse, for set production. Nothing arrived with the RCA nameplate until 1923, when a two-tube regenerative Radiola made its appearance.

The assignment of transmitter manufacture to AT&T was soon forgotten by the other partners, and AT&T charged that the G.E.–Langmuir high-vacuum patent on the triode was an infringement of its own Arnold patent. Both patents were later thrown out by the Supreme Court as involving no invention, in one of that Court's several legalistic, technically questionable decisions.

The manufacture of receiving sets supported several broadcast stations, such as KDKA and WJZ of the Westinghouse Company, WGY of General Electric, WEAF of AT&T in New York, and WLW of Crosley in Cincinnati. Operations were financed as advertising for the companies' products. Then AT&T started accepting general advertising for broadcast by WEAF and the decision for the future support of radio broadcasting had been made. As a theme of his 1930 IRE presidency, deForest pleaded for elimination of advertising on the radio waves but he was too late to reverse the flow of gold from broadcast advertising.

## THREE-HANDED AMERICANS: THE NEUTRODYNE

To increase the ability of their receivers to pick up long-distance signals, some manufacturers included a stage of am-

*Alan Hazeltine with his Neutrodyne receiver.*

plification ahead of the regenerative detector circuit. This made the circuit adjustment even more critical than in the regenerative detector alone, as internal feedback of signal allowed the first amplifier to become a potential oscillator, producing more receiver squeals.

R. V. L. Hartley of the Bell research group in 1918, and C. W. Rice of General Electric in 1920, had proposed circuits which partially balanced out the feedback. In 1923, Prof. Alan Hazeltine of Stevens Institute of Technology produced a circuit with an extra capacitor and two coils, which could completely stabilize a radio frequency amplifier. The circuit was usually designed with three tuned circuits and three dials, including two stages of radio frequency (RF) amplification and a tuned detector. Since a nonregenerative detector was used, the circuit was free of patent infringement.

To reduce magnetic coupling between the tuning coils, the three coils were assembled at a critical angle, mathematically derived by Hazeltine as 54.7°. The three dials and the coils at this angle identified a "Neutrodyne" receiver. The first sets were produced by Freed–Eisemann Radio Corporation in 1923, and soon there were a number of licensees paying royalties to Hazeltine. He founded a research and consulting service, and was President of the IRE in 1936.

The press commented on the necessity for "three-handedness" in tuning the Neutrodyne, since the three dials were supposed to be adjusted simultaneously. The first mechanical solution that reduced the three dials to one was in a Mohawk Neutrodyne of 1925; this placed the three tuning capacitors on one shaft. Atwater Kent produced a set without a cabinet with the three tuning capacitors linked by metal belts. At the time, John V. L. Hogan held a patent, dated 1912, with several years more to run which broadly specified the conditions under which two or more circuits might be tuned simultaneously. Hogan licensed 40 to 50 companies and received a fair royalty income. Fritz Lowenstein held a more narrowly specified patent on the same subject. These two early patents had been shots made in the dark ages of radio.

# 200 METERS AND DOWN

Marconi had incorrectly assumed that only long waves were useful for radio transmission over long distances, and so the regulations of 1916 had consigned the amateur operators to "200 meters and down," an apparently worthless portion of the radio spectrum. As a means of overcoming the distance limitation, the amateur operators had organized the American Radio Relay League in 1914. First used as a means of charting message relay routes across the continent, the League later became the amateurs' technical leader and was ultimately their protector against governmental encroachment.

With improved equipment, the relay routes became longer, but the Atlantic Ocean remained a challenge to be met without benefit of relays. The League arranged a special series of tests in 1921, in which 30 American stations were heard in Europe and one French and two British stations were heard in the U.S. By moving down further into the "worthless" wavelengths, in 1923 two American operators, Fred Schnell (whose call sign was 1 MO) and John Reinartz (1 XAM), connected in a two-way exchange with Leon Deloy (8 AB) in France at a wavelength of 110 m. This feat was accomplished with a power of less than 1 kW.

Soon the wavelengths at 20 m were found useful for daylight long-distance operations. Thus, it turned out that the short waves were better for long-distance radio than the longer waves, and the Commerce Department assignment of the amateurs to wavelengths below 200 m was shown to have been an historic technological mistake.

International bedlam ensued on the short waves until an international conference in 1924 partitioned the short-wave spectrum to the various services. Amateurs were given bands of frequencies as far down as 5 m.

Commercial interests were not idle in those years. For example, a 20-kW tube was produced, and six of them were used in a transmitter that outperformed the old Alexanderson alternator.

# ARMSTRONG AND
# SUPERREGENERATION

At about this time, Armstrong announced his third invention, by which regeneration was carried to an extreme and one tube could produce an output direct to the loudspeaker. He called it "superregeneration." As a business policy, RCA refused to pay royalties and therefore had been excluded from the use of the Neutrodyne circuit. Seeing the need for a new circuit to boost receiver sales, Sarnoff was interested in Armstrong's new invention. For the first time, Armstrong prepared well for negotiations. For the superregeneration patents he received $200 000 and 60 000 shares of RCA stock; the stock was later to make him wealthy.

*The U.S. Army Signal Corps Type BC-127 radio telegraph transmitter of 1924.*

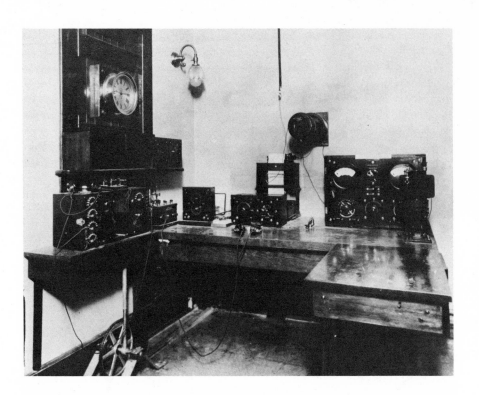

*The receiving room of station NAA in Arlington, VA, in 1915.*

However, the superregenerative circuit lacked the selectivity needed for the crowded broadcast frequencies. This made it suitable only for wide-band high-frequency channels. The invention was used for "identification of friend from foe" circuits during World War II. Otherwise it remains on the shelf, awaiting some new application.

# RADIO ENTERS
# THE LIVING ROOM

The early radio receivers were typically assembled on a wooden plate behind a black bakelite panel, surrounded by a utilitarian wood cabinet. The Neutrodyne, with its three dials, required a horizontal cabinet, on top of which the loudspeaker (which was always sold separately) could be placed. The advent of single-dial tuning allowed more flexibility in design, and as radio entered the living room its cabinet became more ornate. It had to be a piece of furniture acceptable to the housewife.

At that time, the plate or "B" voltage for the tubes was supplied by dry batteries in blocks of 45 or 90 V, connected by wires behind the radio set. The filament battery current was supplied by a lead-acid 6-V storage battery concealed under the radio table. In 1923, the type 201A triode was introduced. It required 0.25 A at 5 V for its thoriated-tungsten filament, and replaced the older 201 tube with tungsten filament, which needed 1 A. Less frequent charging of the battery was needed, but the hazard of acid spills on the living room carpet was not reduced.

The next step was to place the power source inside the cabinet and supply the set from the household socket. It was not hard to replace the "B" battery with a transformer, rectifier diode tube, and inductance-capacitance filter circuit to smooth the pulsations in the rectifier output. But the heavy storage "A" battery could be replaced only when a new tube was designed. First came the type 226, with a heavy high-current filament whose thermal inertia was intended to maintain the temperature and electron emission throughout the ac cycle. This tube was not wholly successful and shortly was replaced with the type 227, the first of the so-called heater-cathode tubes. Electron emission came from an oxide-coated metal sleeve (the cathode) surrounding a tungsten wire heater element. This design had the inertia necessary to maintain temperature and emission. Using it, all-ac sets became available in 1926, enclosed in more attractive cabinets. The loudspeaker, now based on the moving-coil design, was installed behind a decorative grille.

In 1924, RCA used the superheterodyne circuit in a two-dial battery-powered model, and in 1927 brought out a similar all-ac set. Others started producing superheterodynes, and RCA, recognizing the realities of the situation, opened the circuit patents for general licensing in 1930. In England the more complex Neutrodyne and superheterodyne were not generally used because of their high cost. Crystal sets persisted until near the end of the 1920's.

Another of radio's frequent revolutions occurred in 1927, when A. W. Hull of the G.E. Research Laboratory reasoned that the best place to overcome the oscillatory tendency of the triode tube was within the tube itself. He introduced a second grid between control grid and plate to serve as a grounded screen. This was the tetrode screen-grid tube. The troublesome feedback of energy from plate to grid was stopped; higher gain and stable RF amplifiers were then possible without neutralization. This made royalty payments to Hazeltine unnecessary and sales of Neutrodyne receivers essentially stopped by 1930, as tetrode-equipped sets became available. Walter Schottky in Germany also introduced a second grid to produce a linear audio tube capable of increased gain and power output. Still other grids were added for special purposes. These multigrid tubes included the pentode, hexode, and heptode. Later, two or more individual tubes were built in one vacuum envelope, to save space.

The problem of aligning the oscillator circuit and the radio-frequency circuit at different frequencies for single-dial operation of the superheterodyne was solved and that circuit became the standard design.

During the depression of the 1930's, competition among radio manufacturers — Philco, Zenith, G.E., Stromberg Carlson, RCA, Emerson and others — sharpened, and the major share of the limited market went to the independent companies. Sales gravitated more to small portable ac receivers, one popular model being a five-tube superheterodyne in a plastic case.

Armstrong objected to the reduction of amplifier stages in the super-heterodyne. He felt that it debased the reception quality possible with his circuit. Design was essentially frozen, and there was dissatisfaction with the treatment of engineers, since their work seemed to consist largely of finding a five-cent capacitor to replace a seven-cent one. Radio set design was on a plateau of mature mediocrity because cost reduction had become paramount.

# ANTITRUST ACTIONS

The RCA agreement was revised in 1930. AT&T retained rights to the use of tubes in the communication services and sold station WEAF to RCA. That year marked the end of the ten-year prohibition on manufacturing, and RCA took over manufacturing plants in Camden and Harrison, NJ, and proceeded to build their own equipment.

To profit from broadcast programming as well as to foster the equipment portion of the radio business, Sarnoff started the National Broadcasting Company. Thus were born the Blue and Red networks, each a chain of affiliated stations transmitting programs and advertising received from a central source by telephone line. Belatedly, the government determined that this scheme might lead to some violation of antitrust statutes, and the U.S. Justice Department forced divestiture of one chain, which became today's American Broadcasting Company chain. The Columbia Phonograph Record Company, fading because of radio competition, was used to set up the Columbia Broadcast System.

The government also secured a consent decree under which G.E. and Westinghouse bowed out of RCA. RCA now could manufacture its own equipment for sale, and it also owned the major network for programming to encourage radio's use. Thus it could profit from radio's advertising and collect royalties from the use of many of its 4000 patents.

# ARMSTRONG
# AND FREQUENCY
# MODULATION

Edwin Armstrong had another surprise for the field in the 1930's. He developed a way to exploit a new approach to signal modulation — frequency modulation (FM). Radio signals must be of high frequency in order to radiate efficiently; music and speech occur at the much lower audio frequencies, below 15 kHz. In the broadcast signal, the music and radio frequency must be joined; that is, the radio frequency is varied (modulated) by the audio frequencies. This variation can occur in any of three carrier wave parameters: amplitude, frequency, or phase. The easiest form of modulation is to alter the carrier wave's power (amplitude), so the radio pioneers had chosen amplitude modulation (AM). In AM, the strength of the radiated signal is dependent instant by instant on the amplitude of the audio signal. FM had been neglected as an alternative; in fact, it had been unjustly condemned.

In 1922, there were over 300 broadcast stations jammed into a narrow frequency band, and a search was on for a method to narrow the frequency band taken by each station. In AM, the band is twice the range covered by the original speech or music. In practice, this is limited to ±5 kHz on each side of the center or carrier frequency. In 1922, John R. Carson of the Bell engineering group wrote an IRE paper that discussed modulation mathematically. He showed that FM could not reduce the station bandwidth to less than twice the frequency range of the audio signal. Since FM could not be used to narrow the transmitted band, it was not useful. Thus was the kiss of death given to FM.

Actually, Carson had reasoned only that narrow-band FM would distort. Armstrong took the opposite view and expanded the signal bandwidth to 200 kHz and found a useful result. But in 1935, when many hundreds of broadcast stations were jammed into a 1000-kHz band, any proposal to use more spectrum space for each station was heresy. Use of a 200-kHz channel called for moving to higher frequencies, and Armstrong took his experiments to 41 MHz.

Two problems that had been with radio from the beginning were atmospheric static and man-made electrical noise. This interference contained large variations in signal amplitude. Armstrong proposed that frequency variations should carry the useful signal. Amplitude variations could then be stripped off the FM signal, largely eliminating noise and static. Reception highly faithful to the input signal was to be expected. The era of "high fidelity" had dawned.

A circuit was needed to vary the frequency of the transmitter in accordance with the microphone's audio signal. The transmitter could operate constantly at peak output, that is, no reserve of power was needed as in AM, in which a 25-kW broadcast signal requires a transmitter capable of reaching 100-kW on peaks. FM required a "limiter" to strip off all amplitude noise in the receiver and a detector to convert frequency variation into amplitude variation. This would make the signal ready for audio amplification and the loudspeaker.

This system was built by Armstrong on tables in a basement at Columbia University and demonstrated to Sarnoff in late 1933; Armstrong had agreed to give RCA first chance at future inventions. Field tests ensued, with a transmitter on the Empire State Building. These showed freedom from interference to a range of 125 km (80 mi). FM was the topic of a paper and demonstration by Armstrong before the IRE in New York, but not until 1935 (the field tests were the reason for the delay in presenting the paper).

# ARMSTRONG BATTLES
# FOR HIS LIFE —
# AND LOSES

During the next years, Armstrong battled to improve his system and to sell it. In 1938, General Electric asked for a license to produce equipment under the FM patents. This action showed the influence of W. R. G. Baker, head of G.E.'s electronics department, who was enthusiastic about FM. Armstrong's own experimental station at Alpine, NJ, went on the air in 1938, after an expenditure of $300 000. A network of New England FM stations developed (the Yankee Network), which used relays located on mountain tops. Later, the sound channel of the television signal was transmitted by FM. RCA remained aloof, protecting its investment in AM stations and networks. In 1940, RCA made an offer to Armstrong for a royalty-free license for FM, which he did not accept. Armstrong felt that if FM were allowed to challenge the established AM field on an equal basis, FM would supplant AM.

In 1946, RCA announced a limiter-discriminator circuit to circumvent Armstrong's patents, but it was a circuit in which quality was subordinated to

*Daniel Noble with a component from an early FM mobile two-way set.*

lower cost. By 1949, when the basic patents on FM had only two years to run, Armstrong sued RCA and NBC for infringement. This suit was based on his desire to protect his invention from cost-cutting and inferior design, as he saw it. The suit was to drag on for five years of pretrial depositions and cross-examination, much of it trivial. Armstrong was in the witness chair for a solid year.

In the meantime, FM had a triumph with its application to police and mobile radio communication by Daniel Noble of the University of Connecticut. In mobile radio, the noise-combating property of FM was a critical benefit. Noble went on to employment with Motorola, and that company entered the mobile radio field. In 1939, FM was used by Bell Laboratories in a radio altimeter that used signal reflections from the surface of the earth.

RCA proposed a settlement of the suit under terms which Armstrong again rejected in late 1953. In that winter Armstrong apparently suffered a brain ailment, and his financial resources were drained by the Alpine station and his legal fees, but the suit dragged on. His attorneys assured him that a settlement with RCA could be reached, but he was pessimistic. His disturbed mental attitude brought an estrangement from his wife, whom he had met when she was Sarnoff's secretary in 1923. On the night of January 31, 1954, he wrote a letter to his wife. Dressed in hat, overcoat, scarf, and gloves, his body was discovered the next morning on the roof of a third-floor extension of his apartment building in New York.

It has been said of Armstrong that as a player in the game of life, in four at-bats he succeeded in hitting three home runs and received a base on balls. This seems sufficient to justify long-lasting recognition and remembrance from a grateful world. Today in the U.S. there are more stations broadcasting in FM stereo than in AM.

# HALF A SIGNAL
# AND MORE

During the 1930's, the telephone interests, led by Bell Laboratories, continually foresaw new demands for their services. Carson's modulation analysis had shown that an AM wave was composed of a central or carrier frequency of constant amplitude and two side frequencies for each frequency component in the original audio signal. Non-mathematically oriented engineers debated for many years whether the sidebands actually existed or were "mere mathematical fiction." Today we can "see" the sidebands on a cathode-ray screen.

One set of sidebands is above the carrier frequency and the other is below it by a like amount. Since each sideband contains the whole signal intelligence, Carson reasoned that one sideband could be dispensed with, reducing the frequency space for an AM signal by a factor of two. He also found that the carrier frequency, the major power-using component of an AM signal, need not be transmitted and could be supplied at the receiver. Thus, we have single-sideband (SSB) transmission in a narrower frequency band and with much less power required at the transmitter.

The system was put to use in transatlantic telephone service and in carrier-current wire telephony where it doubled the number of channels per pair of wires. After World War II, SSB was taken up by the radio amateurs who needed more operating space in their crowded bands. At first it was thought to be a technique only for skilled operators, but stable oscillators were developed, and the system is now widely used for world-wide communications.

# MUSINGS
# ON A FERRYBOAT

In 1927, Harold S. Black, commuting to Bell Laboratories on a Hudson River ferryboat, scratched out the solution to a long-standing problem on a sheet of the *New York Times*. Thus was invented "negative feedback." By this invention, the output signal of an amplifier is fed back and compared with the input signal, and a "difference signal" developed if the two signals are not alike. This signal, a measure of the error in amplification, is then applied as additional input to correct the functioning of the amplifier, so as to reduce the error signal toward zero. When the error signal is reduced to zero, the output corresponds to the input and no distortion has been introduced.

The invention has been applied in many ways, and the principle remains a cornerstone of electronic circuitry. It was especially valuable for the Bell System's buried telephone cables, which had many repeater amplifiers. Cumulative distortion would have reached intolerable levels without correction of each amplifier by feedback.

The principle of negative feedback was old in mechanical usage. One of the better-known applications was James Watt's centrifugal governor for the steam engine. Two pivoted weights, rotated by the engine shaft, measured the engine speed and fed back a signal to the steam valve to control the input steam and, thus, the engine speed. Black showed how to apply negative feedback to an amplifier circuit and how to predict feedback performance mathematically.

Negative feedback differs in its effects from the positive feedback used by Armstrong in 1913. In the negative-feedback amplifier, gain is reduced and stabilized, whereas positive feedback introduces the tendency to oscillate, with uncontrolled gain.

Harry Nyquist of Bell Laboratories further developed understanding of the negative feedback circuit. He gave us the Nyquist criterion, which predicts the allowable limits of gain and internal phase shift. This solved a problem that had plagued circuit designers.

# BELL MEETS THE
# COMMUNICATION NEEDS

Lloyd Espenscheid and H. A. Affel of Bell Laboratories showed in 1929 that coaxial cable could carry the wider frequency bands needed to transmit more voice channels in telephone transmission. In 1942, such a cable was laid across the continent, providing 600 voice channels and using 600 repeaters, one every 8 km (5 mi). For the needs of long-distance transmission of television signals after World War II, a radio network was designed with horn radiators on towers across the countryside

*Left: TELSTAR, launched in 1962. Right: Glass fibers, descendants of the iron telegraph wire. A typical fiber-optic cable will contain over 100 such fibers and carry over 40 000 voice channels.*

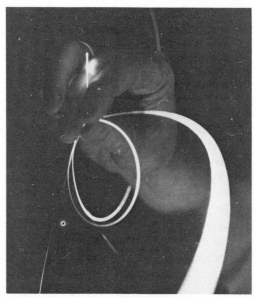

*The Edison phonograph in its most basic form.*

using microwave frequencies around 4.2 GHz. These relays could carry five TV channels, with a sixth available as a spare.

As long-distance calls increased, the Bell System introduced long-distance direct dialing in 1951, and this has now become a world-wide practice. In 1956, after a large amount of engineering work, a transatlantic telephone cable was laid. Repeater amplifiers were installed directly in the cable structure, and were designed for 20 years of service. In 1960, to demonstrate the capabilities of satellites for increasing the number of international channels (see Chapter 8), Bell bounced signals across the continent with ECHO, an orbiting plastic balloon with a radio-reflective surface. In 1962, TELSTAR was launched as a satellite with active transmitting equipment powered by solar cells. Direct-dial circuits from the U.S. to Australia via satellite now make connections in a few seconds.

To obtain still-wider frequency bands and increased channel capacity, the telephone system is currently using infrared frequencies for transmission over tiny glass fibers. The system uses lasers (light amplifiers) and light-responsive solid-state diodes. This system will reduce the size of cables and allow more line capacity to be installed below our city streets. The use of light as a medium also eliminates electromagnetic interference, and this will be an important feature of industrial electronic applications in the future.

# MORE MUSIC
# IN THE HOME

Interwoven through these years and profiting from the technical advances in audio signal amplification, has been the development of the phonograph. As invented by Thomas Edison in 1877, it consisted of a tinfoil-wrapped cylinder revolved by hand. A steel needle contacted the tinfoil, and sound energy transmitted from a diaphragm at the small end of a horn inscribed a sound track in the tinfoil. The movement in the track was up and down ("hill-and-dale") on the tinfoil. Upon again rotating the cylinder, the needle translated the indentations in the track into vibrations of the diaphragm and the recorded sound could be heard from the horn. Crude but intelligible records were obtained.

A major difficulty was that the tinfoil could not be removed from the cylinder without destruction, so the entire cylinder had to be stored. This problem was overcome by Emile Berliner of Germany, who between 1887 and 1894 developed and brought to market the Gramophone with a flat disc and a record player of the same form as today. During playback, the needle operated a diaphragm at the base of a morning-glory horn to channel the sound. Since both recording and playback involved only acoustic energy, the output was weak and the tone was reedy. The necessity for singers and orchestral players to crowd around the horn made recording of large groups difficult.

The Western Electric group demonstrated an electrical recording and reproducing system, the "Orthophonic" phonograph, in 1925, and Goldsmith reported development work on a "new radiola phonograph" in the RCA laboratory. Some of the Western Electric methods were quite sophisticated. Engineers translated the mechanical and sonic elements of the system into equivalent electric circuit elements, and then studied the performance of the reproducing equipment by mathematical solution of these analog electrical circuits. For the phonograph, bandwidth was increased from the previous limits of 300 Hz–2.5 kHz to 125 Hz–4 kHz. Western Electric made their design data available on a royalty basis.

At General Electric, C. W. Rice and E. W. Kellogg used similar methods to design a cone-loaded dynamic loudspeaker. In 1940, D. B. Parkinson and C. A. Lovell of Bell Laboratories carried this method further, and employed electrical circuits and amplifiers to solve differential equations in an analog computer. The first application of this computer was in control of antiaircraft guns in England during World War II. These design triumphs demonstrated the practical uses of circuit theory as the mathematical abilities of electrical engineers improved.

# WAR OF THE DISKS

The early 78-rpm shellac phonograph disk played for only 3–5 min. After World War II, Peter Goldmark of the CBS laboratories introduced the long-playing (LP) record — 33⅓ rpm, 30 cm (12 in) diameter, and made of a quiet vinyl chloride material. Smaller grooves were used and reliance was placed on the electronic amplifiers to raise the output level. The record played for about 25 minutes, so that with a suitable record changer, a lengthy concert could be enjoyed with minimum interruption.

In disk recording, there is danger of overcutting the grooves at low frequencies, where the signal energy is greatest. To avoid this, the bass signals below 500 Hz are reduced in amplitude by resistance–capacitance circuits, and a corresponding boost is employed in the reproducing amplifier to restore the bass frequencies to their original level. At high frequencies, there is room in the groove for larger excursions of the cutting head, and above 2100 Hz the response is boosted to drive the cutting head over a wider range. This is corrected in reproduction by a circuit that reduces the high frequencies to the original level.

Most of the noise reproduced from the record surface appears in the high-frequency range, so when the high-frequency signals are reduced by the correcting circuits in the reproducer amplifier, the noise from the surface of the record is also reduced. Automatic circuits to enhance this effect are built into some amplifiers, notably the Dolby circuits, devised by R. M. Dolby of Ampex.

Shortly after the LP record was introduced, RCA brought out an 18-cm (7-in) record to play at 45 rpm. This had a playing time of about 5 min. It had a large center hole that fitted on a large-diameter center post on a small record changer. This action by RCA was viewed as an attempt to corner a portion of the recording market, and it forced the 78-rpm disk out of use, since the playing time was little different and the new record was smaller and lighter. Eventually, the 45-rpm record became standard for popular single numbers and the LP became the medium for albums, concerts, and high-fidelity recordings.

Stereo sound on records was introduced in 1958. The present Westrex standard employs two tracks in one groove. Movement of the cutting head and the playback needle occurs on two axes at right angles to one another. On one axis (known as L + R), the two signals generated from the L (left) and R (right) microphones are added; their difference controls the other axis. In the amplifier, these signals are processed to produce separate L and R signals. Direct addition of the two signals gives monaural output for nonstereo amplifiers.

Stereo sound was provided for in FM broadcast service about 1961. The sum of the left and right signals modulates the main signal, and the difference modulates a subcarrier displaced 19-kHz from the main carrier. Processing of the signals occurs in the receiver. Nonstereo receivers respond only to the main carrier.

# WIRE AND TAPE

Steel wire was used for magnetic recording by Poulsen, a Dane, in 1898. The recording head was a small coil on a magnetic core, and was held against the wire with a tiny air gap. Magnetized regions were produced in the wire, and the signal was reproduced by a similar reading head. Unfortunately, the wire could roll and move the magnetized area away from the head. Use of a thin plastic tape coated with a magnetizable iron

*Left: Bells Labs developed the orthophonic electrically cut wax record. This 1929 photograph shows the equipment used.*
*Right: On June 9, 1922, at the University of Illinois, Joseph Tykociner gave the first sound-on-film demonstration. The horn microphone and camera are in front of him; at the far right is the projector; the headphones hanging in front were for the listeners.*

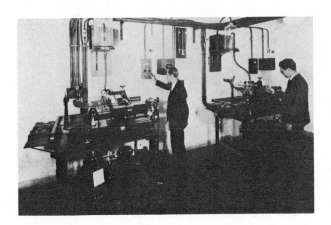
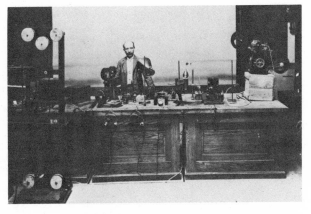

oxide or chromium oxide layer overcame this difficulty in 1948. The use of tape also provided a ready means of cutting and splicing for editing purposes. Although the process is still not well understood, application of a supersonic bias signal along with the recording current was found to improve the recording operation by reducing distortion.

A variety of tape speeds are now used, each speed being half of the next higher one. The speeds in use vary from 76 cm (30 in) per second to 4.75 cm (1⅞ in) per second. Fidelity of the high frequencies is a function of tape speed; 20 kHz is the upper limit for a tape speed of 19 cm/s.

With the recent availability of the digital disk recorder, methods have been devised to utilize its abilities to improve audio recording. Digital modulation in the grooves is "read" by a laser, producing fidelity near perfection, as well as a dynamic power range that covers the range of the actual musical performance. In this system, the audio waveform is sampled at a rate of 44 056 times a second. The sample is encoded in 14-bit binary code, that is, $14 \times 44\,056 = 616\,784$ bits are recorded each second. For stereo signals, this rate is doubled. The digital disk recorder easily accommodates this frequency, since it is designed to record a television signal of up to 6 MHz. In playback, digital-to-analog circuits translate the code to analog form, which is then amplified in a normal manner. The frequency response is flat from dc to 20 kHz. The dynamic range is very large, being that of the sample or $2^{14} = 16\,384$ bits, which represents a power ratio of 85 dB, far greater than the usual tape limit of 60 dB. Thus have techniques from computer and television been combined to yield an important advance in audio.

# PICTURES SPEAK

Moving pictures did not always speak and sing, nor were they always in living color. About 1923, J. T. Tykociner of the University of Illinois, C. A. Hoxie of the General Electric Company, and the Western Electric group all independently saw opportunities to synchronize sound and screen action through the use of electronic techniques and equipment.

The first and simplest method of doing this was by coupling a record player to the moving picture projector. If the record was started when a marked target appeared on the film lead-in strip, the sound would be in step with the film's voice action. Records of 40-cm (16-in) diameter were used to accommodate the running time of a film reel. Careless cuing of the starting point or a needle that jumped a groove could lead to voices out of synchronism with lip movements, and audiences were not tolerant of such errors.

The first moving picture with disk sound, *The Jazz Singer* with Al Jolson, was released in 1927 and had a phenomenal success. Problems arose in the theaters, however, because many were mere boxes without architectural allowance for acoustics. Understanding the screen voices was often difficult in these rooms because of backwall or other reflections. Solutions were sought in draperies, directive loudspeakers, and circuits to remove some of the low frequencies. The filters needed for proper music reproduction unfortunately disrupted voices.

Two other methods of sound recording, involving a sound track recorded directly on the film, were developed to provide automatic synchronization. In the Western Electric system, a track about 2 mm (actually 0.070 in) wide was taken off the side of the picture area, and sound was recorded there by a light valve, producing a variable-density "gray-scale" form of track. This method required a wholly new projector with a sound-reproduction light source that shone through the track to a photoelectric cell. The output of this cell was amplified and delivered to loudspeakers behind the screen which was made of a sound-permeable material.

The variable-density track required care in film processing to maintain linearity between light valve exposure and the density of the silver image on the track. To overcome this, the G.E. Photophone "variable-area" system was marketed by RCA. The sound track, recorded on the film by a mirror galvanometer, was black on clear film, its area varying in proportion to the sound level. Since only black or clear areas were used on the track, the variable-area method was not affected by film processing techniques. Also, since black represented no sound, noise due to film graininess was reduced. These two methods were compatible in the projection machines and soon superseded sound on disks.

# SIX DECADES

At the end of World War I, radio consisted almost entirely of overwater point-to-point telegraph service, a niche in which it could compete with the established wire service. RCA was founded on the premise that the field would continue in that manner; perhaps only David Sarnoff believed that it could do more. With the great inventions of Armstrong, the Bell Laboratories, and many others, radio has become today's electronic audio entertainment industry.

The electron's wonders continue to unfold — colored gold.

## For Further Reading

H. S. Black, "Inventing the negative-feedback amplifier," *IEEE Spectrum,* vol. 14, p. 54, Dec. 1977.

C. Dreher, "His colleagues remember the doctor," *IEEE Spectrum,* vol. 11, p. 32, 1974.

——, *Sarnoff: An American Success.* New York: Quadrangle, 1977.

J. Eargle, *Sound Recording.* New York: Van Nostrand Reinhold, 1976.

A. F. Harrison, "Single-control tuning: An analysis of an innovation," *Tech. Cult.,* vol. 20, p. 296, 1979.

E. Lyons, *David Sarnoff.* New York: Harper and Row, 1966.

O. Mayr, *The Origins of Feedback Control.* Cambridge, MA: M.I.T. Press, 1970.

A. A. McKenzie, "Sarnoff: Controversial pioneer," *IEEE Spectrum,* vol. 9, p. 40, Jan. 1972.

H. Pratt, "Radio ancestors — An autobiography by Robert H. Marriott," *IEEE Spectrum,* vol. 5, p. 52, 1968.

H. A. Wheeler, *Hazeltine the Professor.* Greenlawn, NY: Hazeltine Corp., 1978.

# 6

# THE ELECTRON SHOWS ITS MUSCLE

Central electrical generating stations — the forerunners of today's utility industry — began development soon after the invention of the electric lamp. The number of companies grew rapidly because the low-voltage dc used could only serve small areas. Because of the limitation on transmission distance, many stations were needed to supply cities such as New York or London. By 1900, there were more than 2800 small dc stations in the U.S.

## EARLY CENTRAL STATIONS

Applications of electricity closely followed the availability of reliable and efficient Gramme generators (see Chapter 3). Such machines were used in Paris in 1878 with the Jablochkoff candle, a form of brilliant arc light. Charles F. Brush designed his own generator and lighted the Public Square in Cleveland, OH, in 1879 with carbon-rod arcs. He made an installation in New York in 1880.

Following Edison's development of his incandescent lamp in 1879, with illumination more suited to home or office than was possible with the arc, there were many lighting installations made in homes of the wealthy, complete with

In the 1880's and 1890's, dynamos
(as shown here and on facing page)
were easily identified by their design.
Oil cups (over the bearings) and wide
copper brushes are evident on most of
these, but armature and field magnet
shapes vary widely.
Above left: The Thomson–Houston
dynamo was unique in its spherical
armature, cup-shaped field magnets,
and three-part commutator.
Above right: A dynamo of the
Siemens Company of England.
Right: A dynamo designed by
Charles Brush, with a capacity
of 25 arc lamps.

steam boiler, engine, and generator. In England, Joseph Swan also developed a
carbon incandescent lamp. Both Edison and Swan lamps were shown at the
Paris Exhibition in 1881. After extensive legal maneuvers, the two combined
their interests in 1883 in the Edison and Swan United Electric Light Company
in England.

It was seen by a number of entrepreneurs that one centrally located generating
plant might supply lighting power for a number of contiguous customers.
In 1879, the California Electric Light Company of San Francisco used three
Brush generators to operate arc lamps on the premises of private customers,
later adding equipment when incandescent lamps became available. About the
end of 1881, Calder and Barrett, London contractors, lighted the town of
Godalming, Surrey, England, using a water wheel-driven Siemens ac generator. They started with seven arc lamps and about 40 incandescent lamps, but
when the load had grown only to about 100 lamps by 1884 the service
was discontinued.

Edison, through his London agent, obtained a contract with the City Corporation in January 1882, to light Holborn Viaduct, London, and to supply
private customers. Using an Edison "Jumbo" dynamo, this installation preceded
Pearl Street in New York by eight months. The generator was designed to
supply up to 1000 lamps; underground conductors were used, circuits were

fused, and electrolytic meters were installed. But the relatively low cost of gas illumination discouraged investment in the new technology and the Holborn station ceased operation in 1886.

In February 1882, another installation began operating in England, that of the Hammond Electric Light Company at Brighton, giving service from dusk to 11 P.M. In 1883, the engineer in charge was Arthur Wright, who later was to have an important influence on Samuel Insull's system-building. The Brighton system, in supplying arc lamps and incandescent lamps from the same feeders, required a constant current in the load. This was maintained by employing a boy to watch the needle of an ammeter and adjust a generator field-shunting resistance by the ammeter indications. Arthur Wright later devised a successful automatic regulator to replace the boy.

The Edison generating station in Pearl Street, New York, gave major impetus to the central station industry. Located on the doorstep of Wall Street, Edison's financial fount, it was a well-planned and complete installation serving a neighborhood limited to about 50 city blocks by the low-voltage dc distribution. It employed about 900 hp of reciprocating steam engines driving Edison "Jumbo"

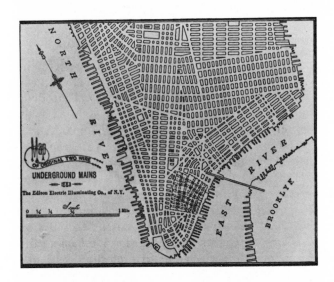

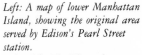

*Left: A map of lower Manhattan Island, showing the original area served by Edison's Pearl Street station.*
*Below left: The Thury dynamo was an English design.*
*Below: The Anglo-American Corporation, although associated with Brush, used dynamos of very different design.*

generators capable of supplying 7200 lamps at 110 V. The initial load in September 1882 was about 1300 lamps on the property of about 60 customers. The station dwarfed previous installations and included such advanced features as mechanical coal and ash handling, steel plate stacks, and direct-connected prime movers.

A considerable change in the operating requirements for a central station came with the increasing use of electrical power in the late 1880's. Dusk-to-dawn operation for lamp loads changed to full 24-hour demand as industrial loads developed.

The 1891 installation of a three-phase ac line at 30 000 V, carrying 100 kW for 160 km (100 mi) between Lauffen and Frankfurt in Germany, created much interest and advanced the ac cause. Central stations, with their dust and smoke, could be located near industrial plants to supply their loads, and transmission lines could supply residential and office lighting and elevator loads elsewhere in the town. Industrial use of electricity developed rapidly after Nikola Tesla, aided by C. F. Scott and B. G. Lamme at Westinghouse, brought the ac induction motor to use in 1892 and Elihu Thomson developed the repulsion-induction motor for G.E. in 1897.

The Niagara development with 25 Hz ac in 1895 showed that it was practical to generate large blocks of ac power and to transmit it to loads elsewhere. In 1896, part of the Falls output was sent over a 32-km (20-mi) transmission line to Buffalo at 11 000 V, three phase, derived from Westinghouse two-phase generators through Scott-T transformer connections.

# DEVELOPMENTS IN EUROPE

In England, the development of central stations faced a number of legal hurdles. By the Public Health Act of 1879, the local governments had control of the subsoil under the streets for water supply, drainage, or public lighting. The most significant obstacle was the Electric Lighting Act of 1882. Prompted mainly by antimonopolistic fears, the act required approval by local authorities for installation and also gave them the right to purchase the undertaking at the value of the physical plant after

*Below: Tesla's 1888 induction motor.*
*Bottom left: This central station interior — the Edison station at Taylor's-on-Schroon in New York State — is typical of the 1880's. Note the pan full of oil cans in the center.*
*Bottom right: The distribution board at the Edison central station in New Brunswick, NJ.*

21 years. This was referred to as the "scrap-iron clause," as it provided no compensation for the value of a going business.

The Board of Trade was the governmental body that regulated the electric light industry. During the years from 1880 to 1885, this body was headed by Joseph Chamberlain,[1] who as Mayor of Birmingham had put the public utilities there under government ownership. As President of the Board of Trade, he viewed government ownership of utilities sympathetically. It was he who introduced the 1882 act in Parliament. In the U.S., Edison had wined and dined New York City aldermen to secure permissions for Pearl Street; in England such efforts had to reach the Speaker of the House of Commons and other prominent M.P.'s. Despite some organized opposition against the compulsory purchase provision, the act passed in April.

By 1883, the effect of the act was evident — 69 provisional grants for central electric stations were approved, of which 62 were later withdrawn. In 1884, the numbers were four issued, four revoked. While the government's right to purchase was considered the major discouraging factor, the general business recession of the time certainly contributed to an unwillingness to invest. Also, gas light was cheaper in England up to about 1910, thanks in part to the invention of the Welsbach Mantle in 1886, which gave gas a competing white light. As long as electric light remained a luxury, the industry was handicapped.

The 1882 act was revised in 1888, with right of government purchase not available for 42 years and the purchase price set at "fair market value." Growth followed, and by 1891 there were 54 central stations in Britain, 17 of which were in London. Prior to 1889, the metropolis was governed by a large number of local authorities, each controlling electricity distribution in its area. After formation of the London County Council, the London area was assigned to seven of the utility companies then operating in the area.

Narrow boundaries fixed the limit of electrical supply, and in 1901 a commit-

---

[1]*The name is familiar in other fields — his older son Austen became Chancellor of the Exchequer in 1903, and his younger son Neville was Prime Minister during the critical years 1937–1940.*

tee of the Institution of Electrical Engineers inquired into the reasons for the lack of electrical development. One spokesman suggested that electrical undertakings should have their boundaries set by economic considerations rather than by arbitrary lines that were mostly of medieval ecclesiastical origin!

One attempt to break out of the traditional pattern in London was made by S. Z. Ferranti. In 1881, he was gaining experience at the Siemens Brothers plant near London; in 1886, at age 22, he made a name for himself at the Grosvenor Gallery Company when he changed their series Gaulard and Gibbs ac system to parallel operation. At the time, this system lighted a large London district, operating with cables strung over the rooftops to avoid the Lighting Act.

In 1889, The London Electric Supply Company chose Ferranti to plan a new station at Deptford on the Thames. Ferranti planned for two 1250-hp dc generators and four 10 000-hp engines driving 10 000-V single-phase alternators. This was a huge installation, and the London press gave Ferranti the stature of an Edison. Transmission of power to London was to be at 10 000 V, by cable buried along the railway lines (in order to avoid street excavation and conflict with the Lighting Act).

The first power was transmitted to London from the 1250-hp machines in 1890, but the station was plagued by problems with the generators, boilers, and coal-handling equipment, possibly a result of taking too large a first step or of an overly optimistic attitude by the financiers. When Ferranti learned that the 10 000-hp engines would not be installed, he resigned. He had correctly foreseen the future in locating the plant outside London, but he underestimated the importance of polyphase ac in building the industrial motor load. London had chosen dc and single-phase ac, and by 1891 was becoming well lighted but still was not in the industrial mainstream.

During the period of London's troubles, Charles Merz and his father J. Theodore Merz had built a major utility system based on the Newcastle on Tyne Electric Supply Company (NESCO), supplying the Newcastle area at 40 Hz. The region was favored with local coal, which had attracted industry, thus providing an industrial load for the electrical system. The leaders of this community were organized by Charles Merz and used their political and financial power to support the privately owned NESCO. Thus, the company was successful in securing passage of several parliamentary bills favorable to its expansion.

Merz stressed the regional characteristics of the system. He was particularly interested in the use of the waste heat from the coke ovens in his stations. NESCO fought inclusion in the National Grid plan, and the change of frequency to 50 Hz entailed problems, but both changes were made and ultimately paid for by the government. The achievements of Charles Merz and his associates showed that legal obstacles placed on system development could be overcome.

Merger attempts among British utilities met with municipal jealousies and incompatible voltages and frequencies. The technical differences made modification difficult while carrying the load of World War I. As late as 1924, it was reported that within London there were 17 different frequencies employed. Finally, in 1930, the country undertook an expensive frequency standardization, settling on 50 Hz, by then common in Europe.

After World War I, another investigating committee concluded that electrical development had been unsound. Government ownership or control became a central issue, and on April 1, 1948, the government took over the utility industry, with payments at market prices made to the stockholders.

In Germany, the Lauffen-to-Frankfurt ac line of 1891 was an important element in the decision of the Frankfurt city authorities on the ac versus dc question. Cologne had already chosen ac. The Frankfurt Electrotechnical Exhibition allowed the various vendors to demonstrate their equipment. Oerlikon of Switzerland and Allgemeine Elektrizitäts-Gesellschaft (AEG) showed that their oil-immersed transformers could safely operate on the Lauffen line at 40 000 V.

Early improvements in incandescent lamps came from Germany. Walter Nernst of Gottingen developed the Nernst lamp, a light source of heated refractory oxides, in 1899. Siemens and Halske patented a tantalum filament lamp in 1902, and Carl Welsbach produced the osmium filament lamp. These challenges to the lucrative carbon lamp business alerted Steinmetz and General Electric at Schenectady, and the G.E. Research Laboratory was established in 1901, with Leipzig-trained Willis Whitney in charge.

Berliner Elektricitäts-Werke (BEW) was the largest of the German utilities. Owned by AEG, in 1895 it had four central stations, all dc, with a combined capacity of about 9900 kW. In the early (1884) negotiations, the Berlin city administration had noted the depressing effect of Britain's 21-year repurchase agreements, and agreed to a 30-year interval. The city conceded a monopoly on a central Berlin area, suited to contemporary dc distribution. The 1891 ac demonstrations made that agreement obsolete, and prolonged negotiations led in 1899 to a renegotiated BEW franchise covering all of Berlin. The installed capacity was to be limited to ensure that the rates did not reflect too much idle capacity.

By World War I, BEW had a uniform supply system for Berlin. In 1915, the City purchased BEW at book value, as agreed upon in 1899. On the whole, per capita consumption was double that in England, and the development of an ac transmission net had proceeded well. There was close integration with the major German manufacturers, AEG and Siemens.

Industrial progress in both the U.S. and Germany was aided by the fact that governmental dealings occurred only at the local level, not at the national level as in Britain. It is also evident that in the U.S. and Germany new equipment was constantly made available to the industry by major domestic manufacturers. In Britain, no domestic manufacturer dominated the market, and consulting engineers and utilities tended to specify equipment in the diverse patterns of the past.

# ELECTRIC TRANSIT

Horse-drawn passenger carriages, operating on steel rails, appeared on the New York streets in 1832. Boston had them in 1836, and Paris and London followed in the 1850's. Steam power was used in some cities, but was largely rejected because of the smoke and dirt of coal.

Werner von Siemens built the first electric street railway at Lichterfelde, near Berlin, in 1881. By 1887, there were 21 electric street railways in operation in North America, the most publicized ones being in Boston and Richmond, VA.

The latter line was installed by Frank Sprague and employed 40 cars on a hilly, curved, 20-km (12-mi) line.

The geared series dc motor was found to be ideal for such service because it had maximum torque at low speeds, which was needed to start the cars on hills. But a major motor problem appeared in the Sprague operation in Richmond. At that time brass blocks were used to contact the commutators, and because of sparking on the hills both block and commutator required filing and smoothing at almost every passage of the car past the repair yard. In 1888, Charles Van Depoele of the Thomson–Houston Company suggested the use of carbon as a brush material. These brushes were tried and found to provide a polished commutator surface without undue wear of the carbon. Frank Sprague also developed the multiple-unit control system, whereby each car in a train could have its motors under control of the motorman in the lead car.

Power supply was usually at 500–600 V dc from an overhead wire on which ran a contact shoe or trolley, with current return through the rails. Where a third rail could be protected from accidental contact, it was used (as in the subways). In 1895, electric power was first applied to a main-line railroad to eliminate the smoke and heat from steam locomotives in the Baltimore and Ohio railroad tunnel in Baltimore. Also in that year, a branch line of New York, New Haven and Hartford was electrified, followed in 1907 by complete electrification of that railroad's system.

Subway systems, originally used with steam locomotives but soon electrified, continue to be used and new ones built in major cities, but the era of the street railway passed within 60 years due to extensive use of the automobile and the motorbus. The street railways in the U.S. had supplied a major daytime load for the early central stations. This was not so true in London, where the railways and tram lines had built their own generating stations.

# PLANTS AND
# GENERATORS GROW

The reciprocating steam engine, usually direct-connected to the generator, reached 5000 hp by 1900. The 10 000-hp engines built for Ferranti's Deptford Station required the casting of the heaviest steel ingots in the history of British steelmaking, indicating that a much greater engine size might not be possible.

Fundamentally, the reciprocating engine and the electric generator were a mechanical mismatch. The reciprocating engine supplied an intermittent torque, whereas the electrical generator and its load called for a continuously applied torque. This inequality was smoothed by storing rotational energy in large flywheels, but these took up space and were necessarily heavy and costly.

The constant torque of the steam turbine was an obvious answer to the need. Turbine development had been carried on for many years, but it was not until 1883 that de Laval in Sweden designed a practical machine. This operated on the impulse principle, the wheel being struck by high-velocity steam from a nozzle. Rotational speeds as high as 26 000 rpm were produced, but the power output was limited by the gearing needed to reduce the speed to that suitable for the generators. By 1889, de Laval had a turbine which drove a 100-kW generator.

Charles Parsons, an English manufacturer, developed a steam turbine using reaction principles, in which the expansion of steam through a nozzle reacted on the turbine wheel. In 1900, he designed a 1000-kW unit for Elberfeld, Germany. He became associated with Westinghouse where the first U.S. machine was designed. Serving a load of 1500 kW, it was installed in 1901 at Hartford, CT.

Charles Curtis worked with turbines having multiple stages of impulse design and later sold his American patents to G.E. When G.E.'s W. L. R. Emmett,

*Building a dynamo in the late nineteenth century.*
*Left: First, thin disks of iron and insulation were alternately stacked on a shaft to form the armature core.*
*Right: Second, conducting wires were laid around the core (continued on next page).*

*Building a dynamo in the late nineteenth century (continued from previous page).*
*Left: Third, the armature was wrapped to ensure that the wires stayed in place as the armature revolved.*
*Right: Finally, the armature was fitted into the frame holding the field magnets. Design was still hit-or-miss to a large extent, and new armature designs were tested after hours — they occasionally exploded spectacularly.*

working with Curtis, designed a 5000-kW unit for Chicago Edison in 1903, the die had been cast for the future — the industry thenceforth was to be powered by the efficient and economical steam turbine. Within the next year, G.E. and Westinghouse received orders for turbogenerators totaling 540 000 kW load capacity.

Although not as efficient as originally expected, improvements in turbines have reduced the coal burned per kWh from about 10 pounds at the Pearl Street Station to near 0.7 pounds (0.3 kg) today.

Generator capacity had reached 100 000 kW by 1930. A major increase in capacity then occurred with the installation of a 208 000-kW compound machine at the State Line Station near Chicago. After 1950, further advances in maximum ratings came with the adoption of hydrogen cooling of the generators in a closed system. Windage losses were reduced because of the lower density of the gas and greater heat transfer achieved. In recent years, the cooling fluid — hydrogen or water — has been used in hollow conductors or tubing channels in the slots. These and other improvements allowed designers to reach 1 380 000 kW (1380 MW) in the mid-1970's in a machine at Cumberland, MD.

For energy storage, pumped water is feasible where the plant location is suitable. Reversible hydroturbines are used, operating as motorized pumps to raise water to a lake above the plant at times of excess system capacity. At times of peak demand, these machines are reversed and water is drawn from the reservoir. The concept was first applied in Connecticut in the 1930's. The largest pumped storage installation is at Ludington, MI, capable of 1900 MW peak output. Lake Michigan serves as the lower reservoir; a man-made lake is the upper reservoir. An 1120-MW pumped-storage plant is also in operation in Austria.

# TRANSMISSION LINE
# PROBLEMS

The first large ac transmission line is reputed to have been that installed by Westinghouse in 1891 to connect a waterwheel generator of 100 hp to the Gold King Ore Mill at Telluride, CO. It was operated at 3000-V single-phase over 3.6 km (2¼ mi) at an elevation of more than 9000 ft. The need arose from the lack of wood above the timberline and the cost of transporting coal by wagon to that isolated locality. Also in 1891, the Lauffen-to-Frankfurt line was operated at 30 kV. In the U.S., 40 kV was applied on a 110-km (70-mi) line from a remote hydroelectric station to Sacramento, CA, in 1892.

Above 40 000 V, the line insulators posed a limit, being pin-mounted porcelain of considerable size and fragility. In 1907, H.W. Buck of the Niagara Falls Power Company and E.M. Hewlett of G.E. solved the problem with suspension insulators composed of a string of porcelain or glass plates whose length could be increased as the voltage was raised.

Another problem area had been investigated by C.F. Scott and Ralph Mershon (AIEE President, 1912–1913) of Westinghouse. Scott had noted luminosity accompanied by a hiss or crackling sound on experimental lines at night when operating above 20 000 V. Later it was found that radio interference was also produced. The phenomenon was the cause of considerable energy loss, enough to justify further study, which these men undertook on the Telluride line in 1894. They concluded that this "corona loss" was due to ionization of the air at the high field intensities created around small-diameter wires. The power loss appeared to increase rapidly above 50 000 V, posing a potential limit to high-voltage ac transmission.

Scott's 1899 report of the Telluride work interested Harris J. Ryan, then teaching electrical engineering at Cornell. In 1905, Ryan became head of the Department of Electrical Engineering at Stanford, and he continued his high-voltage work there. In the laboratory, he showed that the conductor diameter and spacing might be increased to reduce the field intensity and hence the corona loss. After this work, conductor diameters increased and line voltages reached

*Left: Charles F. Scott.*
*Right: Ralph D. Mershon.*

220 kV by 1923. Hollow cables of 2.5-cm (1-in) diameter were used at 287 kV in 1934 on the 440-km (275-mi) line from Hoover Dam to Los Angeles. In 1970, 765-kV lines were used in the 900-km (560-mi) line from hydroelectric stations on several Arctic rivers to Montreal, and in 1983 a line operating at 1000 kV was planned.

Because the cable weight and consequent tower loading were excessive with large cable diameters (a 6-cm cable weighs about 6 kg/m or 4 pounds/ft), a new approach was developed in Germany during World War II. This used a cable subdivided into several smaller conductors and is now known as a bundle conductor. Two to four small cables are spaced around the periphery of a circle about 30 cm (12 in) in diameter. The electrical stress on the air is reduced to that which would occur if the conductor's actual diameter equaled the assembly's cross section. Although the idea was developed in Germany, the first application was a two-cable bundled conductor on a 380-kV line in Sweden in 1952. Since then, nearly all lines operating at voltages above 300 kV have used bundle construction.

Another line problem was the outages and insulator damage caused by lightning strikes. F. W. Peek, Jr., graduated from Stanford in 1905 and went to G.E. at Schenectady, where he joined the Steinmetz consulting engineering department in 1909. He devoted his life to studies of lightning and methods of avoiding its dangers. Much of his work was done with an impulse generator designed to simulate lightning waveforms and intensities. Thus he brought lightning into the laboratory.

When there are generating stations at both ends of a long ac transmission circuit, there is a problem in maintaining the machines in synchronism. But if high-voltage dc is used, the end stations can operate independently and very long lines become practical. For very large blocks of power, the additional cost of the ac-dc-ac conversion equipment (mercury-arc and solid-state rectifiers and inverters) can be absorbed and the use of high-voltage dc becomes economical. In the Soviet Union and the U.S., overhead dc lines operate at 800 kV. In the U.S., this voltage is used in the Pacific Intertie, carrying 1440 MW over 1330 km (825 mi) along the coast from the Northwest to Los Angeles. The rectifiers used were manufactured in Sweden.

Extensive interchange of ac power by interconnected networks began in Europe after 1951, involving Austria, Belgium, France, Italy, Luxemburg, The Netherlands, Switzerland, and West Germany. The initial network operation was at 220 kV, later raised to 380 kV. In the 1960's, underwater cable connected Britain to France, and Sweden to Denmark. The Comecon countries formed similar interconnections, linking Czechoslovakia, Hungary, Poland, Romania, and the Soviet Union.

Questions arising in transmission and generation location were often answered in advance of construction by network simulators or "network analyzers." The first studies took place at M.I.T. in 1929, where Vannevar Bush constructed a line system analyzer with simulated systems of adjustable constants, including lines, transformers, generators, and loads. The manufacturing companies installed more elaborate equipment to enable their customers to set up hypothetical systems. Similar installations of simulation equipment, some operating at times at 400 Hz to reduce inductor losses, were made at Carnegie

*Vannevar Bush with his differential analyzer of 1931, a descendant of the 1929 line system analyzer.*

Tech, Illinois Institute of Technology, Kansas, and other universities. At Iowa State after World War II, a unique analyzer operating at 10 000 Hz was developed to further reduce inductor losses and to permit electronic generators to be used. This design was employed by Yokogawa Electric in Japan and several installations were made worldwide. These simulators have since been replaced by the digital computer.

# BUILDING
## A UTILITY SYSTEM

The invention of the rotary converter—a machine with a dc commutator on one end and ac slip rings on the other—helped to moderate the ac–dc battle, and allowed for an orderly transition of the industry from the era of electric light to the era of electric light and power. Westinghouse used this machine in its universal electrical supply system, based on polyphase ac, which it displayed at the Chicago exposition of 1893. The general use of ac equipment was also aided in 1896 by a patent exchange agreement between General Electric and Westinghouse by which rational technical exchange became possible.

Some utility executives foresaw the need to bring order to the supply of electricity by consolidation of the small stations and their service areas. What happened in the Chicago area from 1892 through the first decades of the twentieth century, as described by Thomas P. Hughes of the University of Pennsylvania, is a good illustration.

In 1892, the year of the consolidation of the Edison General Electric Company and the Thomson–Houston Company, Samuel Insull left the dual position of

Vice President of the new General Electric Company and Manager of its Schenectady works to become President of the Chicago Edison Company. Insull had come from England in 1880 to be Edison's personal secretary, and his method of attacking problems was learned from Edison: it involved a synthesis of technology and economics appropriate to the times, and it did not overlook political necessities. Insull was a leader in utility public relations efforts.

Believing in the economy of large units, Insull's first step in Chicago was to enlarge the system by building the Harrison Street Station in 1894, with 2400 kW of generator capacity. These were driven by reciprocating engines operated in the condensing mode because there was plenty of water available for cooling. By 1904, the plant had been expanded to 16 500 kW, using 3500-kW units. With so much generating capacity available, a large market had to be found, so Chicago Edison started to acquire other companies, ultimately gathering 20 in the Chicago area. Through mergers, it became the Commonwealth Edison Company in 1907.

As small plants were absorbed, the station locations were converted to transformer substations or rotary converter locations. The converter and the motor-generator frequency changer became important elements of planned system growth, since part of the load was at 600 V dc for transit use, part of the industrial load was supplied by generation at 25 Hz, and there was a major load at 60 Hz, which was above the flicker frequency of incandescent lamps. Later, the 25-Hz operation was eliminated. In some cases, the mercury-arc rectifier (invented by Peter Cooper Hewitt in 1902) was employed to obtain dc for transit use.

Insull and his chief electrical engineer, Louis Ferguson (AIEE President, 1908–1909), saw that the space requirements and weight of the reciprocating steam engines at Harrison Street limited the maximum rating for that station. Insull kept abreast of European developments, and had seen a steam turbine in a central station in Germany. Having decided to equip the new Fisk Street Station with turbogenerators, Insull sent Ferguson and Frank Sargent (of the consulting firm of Sargent and Lundy) to Europe in 1901 to inspect turbine installations.

On this trip, Ferguson and Sargent saw much to study. At Milan Edison, the chief engineer informed them that he had placed an order for two steam turbines rated at 4000 hp each. At Brown, Boveri and Company in Switzerland, they learned of an order from Frankfurt for a 4000-hp turbogenerator. They returned, convinced that Europe was leading the way in steam turbines, although Ferguson apparently realized some of the pitfalls when he observed that in Europe there is the individual "design of each machine, the careful coddling of new devices and infinite pains and complication which is of the nature of European engineering," intimating that this care might not follow in American hands.

Insull usually depended on G.E. for his electrical equipment, and he discovered that Charles Coffin, President of that firm, was also interested in steam turbines. G.E. would supply a 1000-kW machine, but because of lack of experience would not give a performance guarantee for a unit as large as 5000 kW, which Insull wanted for Fisk Street. They compromised, with G.E. taking the manufacturing risk and Chicago Edison assuming the installation and building alteration costs. A 5000-kW machine went into service in October 1903.

The unit, vertical in design, was one-tenth the size and one-third the cost of the reciprocating engine–generator initially scheduled for the plant. Although not as efficient as expected, the lower cost and improvements made later units satisfactory. In 1909, the original 5000-kW units at Fisk Street were supplanted by 12 000-kW sets that required virtually no increase in floor space.

The Quarry Street Station in Chicago was built in 1908, and by 1910 had six 14 000-kW turbogenerators with 25–60 Hz frequency changers to allow shifting of load among the four 25-Hz generators and the two 60-Hz generators. The load dispatcher for the system was required to schedule load for the machines of Quarry Street in parallel with Fisk Street's 25-Hz units in order to achieve optimum machine loading. At that time (1910) the consolidated Chicago system had been expanded to 220 000 kW of installed capacity. Today that system has a capacity of nearly 17 000 MW generated by modern units, a 77-fold increase in 73 years.

# THEORIES OF MANAGEMENT

Good management was as important to utility success as was the improved hardware of the stations, and Samuel Insull was a leader in adopting new management principles.

On his 1894 European trip he met Arthur Wright, then the manager of the Brighton municipal station. Wright had been influenced by Dr. John Hopkinson, who was a world-renowned electrical authority. In 1883, Hopkinson had solved the problem of parallel operation of ac generators, showing that it was necessary for the several machines to have similar waveforms, for which the sine wave was assumed although not assured by contemporary generator design. He had also treated the magnetic circuit and had shown Edison's lengthy magnetic circuits to be inefficient. He had communicated ideas to Wright concerning the optimum loading of a system and the scheduling of the customer tariff to cover not only the cost of energy delivered, but also the cost of the capital needed to maintain the system. Wright had developed a metering system measuring not only the use of energy, but also the extent to which each customer used his installed capacity or his maximum demand on the system.

From Arthur Wright, Insull learned of a tariff structure recognizing both energy cost and capital cost to the customer. Load diversity and load factor were additional management principles introduced by Insull before 1914.

Equipment must be optimally loaded to ensure adequate return on the investment. However, electrical load varies with time of day and type of customer: residential load is heavy at night, industrial load during the day, and transit load during morning and evening. If the peaks are diverse, they require less central station generating equipment and transformer capacity than if all the peaks occur at the same time of day, or even of the year. The diversity factor of a station was used in selecting sizes of installed equipment, that factor being the ratio of the sum of the peaks of the separate loads (added as if they occurred simultaneously) to the actual peak load. This factor provided a measure of the amount of equipment needed for a station, equipment on which capital charges were accruing.

Insull's 1914 Chicago customers were divided into several classes according to their load patterns. Had they all called for maximum power simultaneously, the utility load would have been 26 640 kW. On the day in 1914 when the actual system peak occurred (January 6), the demand was only 9770 kW. This showed the diversity in customer demand. The possible maximum far exceeded the actual, and equipment capable of meeting the possible maximum was not needed to carry the system.

Insull also realized the importance of load factor for a station or a system. This was the ratio of the average load to the maximum over a specified period. This figure gave an indication of the efficiency with which the system equipment was being utilized. A customer who took 100 kW for 20 h every day was preferred to one who took 1000 kW for 2 h every day. Each customer used the same energy — 2000 kWh. But the first customer only required 100 kW of generating capacity, whereas the second customer required 1000 kW of capacity that stood idle most of the day while capital charges accrued. Through sales effort and engineering advice, and with power rates that made high peaks uneconomical to the customer, the demand could be leveled, that is, the load factor improved.

To show that his management methods were not limited to major cities, in 1911 Insull integrated isolated utilities serving Lake County, IL, near Chicago. This resulted in the Public Service Company of Northern Illinois. In the process, 55 municipal or privately owned plants were replaced by four large and efficient central stations whose output was distributed over 1400 km (875 mi) of high-voltage lines.

# A CAPITAL-INTENSIVE INDUSTRY

The interconnection of systems did not come without capital. At the time of Pearl Street, money rushed after the electric companies, but by the time of system expansion — 1900–1940 — money had to be sought. S. Z. Mitchell was a man who could find it. He was an 1885 graduate of the U.S. Naval Academy and had worked for the Edison companies in New York. He was later a distributor for Edison products in the northwestern states. Mitchell understood the capital needs of the electrical industry — he said that utilities needed about $50 for every baby born in the U.S.

In 1905, Charles Coffin, head of G.E., and Mitchell devised a financing plan for the almost unmarketable shares of small utilities that G.E. had often acquired in payment for equipment. They organized the Electric Bond & Share Company, which exchanged its stock and bonds for securities held by G.E., and converted those securities into marketable assets by invigorating the issuing companies. This was done through management and engineering aid and further financing for expansion where the project was worthy. It was not a holding company in the sense of control, but it did direct its companies through a fee-based mix of services in management, engineering, and financing.

By 1925, Electric Bond & Share was a de facto holding corporation for five other holding companies. Mitchell and his associates devised a plan by which the operating companies were financed thusly: 60 percent in publicly sold bonds, 20–25 percent in publicly sold preferred stock, and 20–25 percent in common stock reserved for a holding company. By keeping the amount of common stock low, the cost of voting control was decreased. If profits were good, then the return on the holding company investment would be large from common stock dividends. When the operating companies were in the same geographical area, they could be united into systems, improving general operations and exploiting load diversity. When the companies were geographically diverse, financial diversity was also possible, since the separate utilities of a group were not likely to be affected at the same time by natural, economic, or industrial calamity.

Mitchell used an example to illustrate the gains available to the investor in holding-company common stock. For $100 of capital he expected a return of $9. If this $100 was raised through $60 of bonds, $20 of preferred stock and $20 of common stock, then at 6 percent the bonds would require $3.60 and the preferred stock at 7 percent would require $1.40. This would leave $4 for a dividend on the $20 of common stock, a rate of return of 20 percent. Higher returns were obtainable by decreasing the percentage of common stock, which was often retained by the organizers. While the common stockholder took the risk, in the stable electrical market this risk was small. The high returns and the control of fees for services extended to the operating companies were uncovered by the holding-company investigation of the 1930's, which led to the dissolution of the utility holding companies.

# GOVERNMENT REGULATION OR CONTROL?

Monopoly operation of utilities raises the fear in political circles of monopolistic rate abuses, so governments have stepped in to control electric rates. The first U.S. public service commission with rate-setting authority appeared in Wisconsin in 1907. Today all states have such bodies. This legal mechanism is intended to allow a reasonable return on the investment and to protect customers from poor service or unfair charges. The Federal Power Commission (FPC) has been given authority to regulate rates and conditions of service for electricity sold in interstate commerce. The FPC also issues permits for hydroelectric projects in U.S. jurisdictions.

In the early days of the U.S. utility industry, private power companies and municipally owned systems provided power. Since the 1930's, federal agencies, the Department of the Interior and its bureaus, public power districts, and rural electric cooperatives have become additional elements in both distribution and generation.

The utilities in Canada, with abundant waterpower, have developed under provincial commissions, private capital supporting only about 25 percent of the total generation. Using its hydropower, Canada has kept electric rates low, which has led to high per capita consumption.

In Britain, the electric system has been nationalized since 1948. To tie the major generating centers together, a 380-kV supergrid was started in 1963. Nuclear-powered generation in Britain is greater than in any other country.

France nationalized its utility systems in 1948, with large generating stations integrated; per capita consumption is low. Italy nationalized its system in 1963. Hydropower provides about 35 percent of needs, with oil the major fuel source for the remainder; Italy has operated a geothermal plant since 1905. West Germany is a net importer of power, largely from Switzerland, where water-power accounts for 81 percent of capacity.

Ever since the sharp increase in fuel costs caused by OPEC in 1973, U.S. rates and consumption figures have been erratic. In 1915, typical annual residential use was only 260 kWh per household. It had increased to 6000 kWh per year in 1968, due to the great variety of electric appliances and to the heavy load of air conditioning.

Giant Power was a post-World War I U.S. proposal for governmental control, which failed to develop. Morris Cooke, advisor to Pennsylvania's Governor Pinchot, wanted mine-mouth plants built in that state with high-voltage transmission lines sending the power throughout the Northeast. Under control of a Giant Power Board, the transmission companies would purchase power wholesale from the giant generating stations and the existing utilities would become distributors. A bitter pill for the existing power companies was contained in Cooke's proposal: the state would have the right to take over the Giant Power facilities after 50 years. In the "return to normalcy" after World War I, the engineers, including Arthur Kennelly of early AIEE and IRE years, showed that the existing system was working well, and pointed out that government interference elsewhere had retarded the sort of development that Giant Power was intended to promote. The bills were reported negatively to the Pennsylvania legislature in 1926.

# HYDROELECTRIC POWER

Although water mills were known from ancient times, the first important electric generation by water power came at the Lauffen, Germany, and the Telluride, CO, ac installations in 1891, followed by the Niagara installation of 15 000 hp in 1895. A major installation was made in 1913 at Keokuk, IA, on the Mississippi river, including a dam nearly a mile long (1.6 km) with a head of 32 ft (11 m).

After World War I, the federal government built a dam at Muscle Shoals, on the Tennessee River, completing it in 1926. A consultant on this project was Hugh L. Cooper, who later helped the USSR with the design and construction of the Dnieprostroy project on the Dnieper river, dedicated in 1932. The largest Russian hydroelectric development is at Krasnoyarsk and is rated at 6000 MW.

Today, most major hydropower sites in North America have been developed; those remaining are at locations remote from load centers. It is modern practice to locate chemical or metallurgical plants having heavy loads at the hydropower site. Examples are Kitimat in British Columbia and Guri in Venezuela. Among the remote but large developments of recent years are Grand Coulee, the Churchill Falls plant of 5200 MW in eastern Canada, and others ranging from 2000 to 6000-MW rating in the USSR and Venezuela. The largest installation will be at Itaipu, on the Brazil–Paraguay border, to be rated at 12 000 MW when completed.

# GOVERNMENT:
# REGULATOR
# OR COMPETITOR?

The initial decision to build nitrate plants at Muscle Shoals, TN, with Wilson Dam as a power source was made in response to the needs created by World War I. The supply of Chilean nitrates for explosives and fertilizer was uncertain and called for remedial steps, but the project was also weighted with political pressures to spread wartime expenditures around the country and to locate fertilizer projects in the South for restoration of its depleted cotton lands.

The dam was not completed until 1926, and a debate ran on from 1918 to 1933 over what to do with the $100 million facility after the war emergency had passed. There were interests promoting river navigation, as the Shoals had been a barrier to barges; there was interest in flood control, especially after the Mississippi river flood of 1927; and there was also Senator Norris, with

*An aerial view of the immense Itaipu hydroelectric plant, located on the Brazil–Paraguay border.*

*The TVA at work: the Norris Dam under construction.*

his interests in electric power development. But the chief political conflict centered on who was to develop the electrical power — government or private industry — and who was to benefit from the wartime expenditures on the plants and dam. Senator Norris succeeded in passing bills that embodied his belief in public gain through government ownership, but they were vetoed by Presidents Coolidge and Hoover.

In response to a proposal by President Harding that the government lease the project at a rate that would reimburse the Treasury, Henry Ford offered to buy the nitrate plants for $5 million and to lease the dam for 100 years. A pro-Ford bill passed Congress in 1923, but died in Conference Committee, and in October 1924, Ford withdrew his offer.

In 1933, the Great Depression provided the opportunity for several social experiments that would bring the federal government into the electrical industry as a competitor. These experiments were to be carried out through the Rural Electrification Administration (REA), the Bonneville Power Administration (BPA), the Tennessee Valley Authority (TVA), and several other bodies. The TVA was to resolve the impasse over Muscle Shoals, and it also incorporated President Franklin Roosevelt's concept of regional development through cheap electric power.

The REA was to finance rural electric cooperatives, to act as a power wholesaler using federal capital at low interest rates, and to construct distribution lines to serve farms and communities previously thought too small to be economical loads. The Agricultural Extension Services of the land-grant universities showed the farmers what electricity could do for them, and motors rapidly appeared in the barns and refrigerators in the farm kitchens. The private utilities saw in this rural activity an opportunity to build load and many joined in the program, generating the power sold by the cooperatives. Some of the cooperatives have since installed generating facilities. Today, more than 98 percent of all U.S. farms are electrified. The REA is an excellent demonstration of the elasticity of electric power usage.

The BPA and TVA were major components of the Roosevelt program of utility reform, and private utilities reacted negatively to these plans. The TVA competed directly with the Tennessee Electric Power Company, and the issue of the government right to compete with the industry was fought to the Supreme Court. After losing there, Tennessee Power was sold to the TVA in 1939.

The BPA and TVA were initially to build a series of dams along the Columbia and Tennessee rivers to control floods and, by locks at the dams, to provide for river navigation. Hydroelectric power, fertilizer production, and promotion of the region were also objectives. The availability of cheap power attracted electrochemical industries to these regions, and both regions became centers of the atomic effort during World War II.

Roosevelt insisted that TVA establish a "yardstick" for evaluating the efficiency of privately owned electric utilities. But how this was to be done was left unstated. Most of the early power delivered was hydropower and therefore cheap, no taxes were paid, and no dividends or interest were charged to the project.

The yardstick function was approached by TVA through several demonstration communities—towns such as Tupelo, MS, and Athens, AL, where electric rates were reduced to 3 cents for the first 50 kWh, falling to 0.4 cents for anything over 400 kWh per month. The national average at that time was about 5.5 cents per kWh.

Llewellyn Evans, the TVA Chief Electrical Engineer, said, "The rates proposed are designed to encourage and make possible the widest use of electric service, with all the individual and community benefits which go with such wide use." Thus the objective was not to use TVA as a yardstick, but as a tool of social change.

The only chance for success in such a program depended on a sharp rise in per capita electrical consumption, and additional consumption by new businesses, attracted by the rates. In three years, the average rate in Tupelo went down from 7.4 cents per kWh to 1.58 cents and average usage rose from 40 to 178 kWh per month. The customer usage throughout the TVA low-rate area consistently ran above that of the rest of the U.S. By 1941, there was no longer much argument, as the TVA had proved that the demand for power was sufficiently elastic to cover the lower rates by reducing the capital costs per unit of production. The yardstick had become merely a banner at the head of a parade as the investor-owned utilities proved the validity of the load-building philosophy of rate setting.

# THE TVA DILEMMA

In 1940, as the war in Europe developed, the TVA had an installed hydropower capacity of 1 million kW; at the

Table 6.1   PERCENT OF MARKET SHARE

|  | 1900 | 1982 |
| --- | --- | --- |
| COAL | 89.0 | 53 |
| OIL | 4.7 | 7 |
| NATURAL GAS | 3.2 | 14 |
| HYDRO | 3.2 | 14 |
| NUCLEAR | — | 13 |

end of the war this figure was over 2.5 million kW, partly used for aluminum production for aircraft and partly for the demands of the Oak Ridge atomic complex. After the war, instead of a predicted glut of power in the valley, there was a further rise in electrical demand as new industries sought this energy source. By 1948, the river had been completely developed and TVA, facing up to its own creation of massive electrical demand, sought to build steam plants. This move caused controversy in Congress, as the private utility industry fought this further extension of federal power.

By the late 1940's, the arguments for flood control and navigational use of the dams had faded into the background, and TVA fought just to supply the demand created by its early decision to set rates that would build load, a load now threatening to engulf the system. TVA could not meet this demand without steam, so authority was given, leading to 17 million kW of steam installation by 1975; part of this uses nuclear fuel.

The building of steam plants created other problems for TVA. It became the largest consumer of coal in the country, and contributed to the environmental problems of strip mines. Effluent from its stacks was blamed for regional pollution and acid rain throughout the Northeast. The cooling-water discharge of one plant near Paducah, KY, reportedly raised the temperature of the Ohio River 2.2 °C (4 °F).

In the 1960's, the TVA was limited by a requirement that it finance future expansion by the sale of bonds, as do the private utilities; TVA was to operate more on usual business principles.

The rapid rise in fuel costs in the 1970's raised further problems. It demonstrated how success in one era can play havoc with the plans for the next. When the cost of coal increased to over four times the 1970 price of $5 per ton, it was necessary to increase rates that had initially been set on a hydro-generation base. Predictably, the new rates alienated TVA's customers. Demand had been shown to be elastic, a river had been canalized, and a region benefited economically, but at great cost in dollars and in problems for future resolution.

# COAL LOSES A MARKET

Wood has been the primary fuel from the age of the cave man, and coal did not emerge as a competitive source of heat until the sixteenth and seventeenth centuries. The historic London fogs were due to use of soft Newcastle coal, made affordable in London by water transport. Wood was the fuel for the first years of the railroads and steamboats in the U.S., available from the clearing of the forests for early settlements. In fact, wood still constitutes a major energy source in the underdeveloped countries.

In the U.S., coal has had commercial importance since 1822. Colonel Drake's first oil well, in Pennsylvania in 1859, created so much interest that a year later 500 000 barrels were produced. Coal, however, was the fuel of choice from the central station industry's beginnings. By 1970, with coal at $5 per ton and oil at $3 per barrel, the choice of a power plant fuel usually involved only considerations of delivery and storage. The OPEC pricing policies following the 1973–1974 Arabian boycott raised oil prices as high as $36 per barrel and coal moved up to $20 per ton. Plants originally designed for oil were, in many cases, converted to use of coal at great cost.

Despite the higher cost of oil, the demand for coal did not expand as fast as had been foreseen. Coal has lost market share over the years, as shown by Table 6.1.

Coal as mined is a rather intractable dirty rock. Charles A. Berg, formerly Chief Engineer of the Federal Power Commission, has pointed out that coal has suffered because the market did not demand improvement of coal itself. In the use of coal in producing electricity, industry interest seems not to have gone beyond improving combustion efficiency: reduction in pounds of coal per kilowatt-hour has been the only target. Philip Sporn, of the American Electric Power System, once stated that the steady business supplied by the utilities actually made coal insensitive to market needs and made unneccessary any development of a coal science leading to improvement in coal itself. Thus, over the years, coal has become a fuel of last resort. Now it is plagued with environmental problems. In the final analysis, coal will be the fuel of last resort for many decades to come.

# THE NUCLEAR AGE

The nuclear power age began with the reactor at Duquesne Light Company at Shippingport, PA, in 1957. This plant, built by Westinghouse, was designed for 60 MW and used a pressurized water reactor (PWR). Another early successful installation was the Dresden plant of Commonwealth Edison of Chicago, a boiling water reactor (BWR), built by General Electric.

Postwar images of cheap fuel and a rosy future for the nuclear industry persisted. In 1973, it was predicted that one half of the U.S. energy needs would be supplied by atomic fission in the year 2000. But there have been no orders in the U.S. for nuclear plants, and there have been many cancellations since March 1979. This was the date of the disaster at Three Mile Island in Pennsylvania. There, a combination of valve and instrument failure and unprepared operators who misread the warning signals led to massive core damage and ultimate shutdown of reactor No. 2, an accident not yet fully analyzed because evidence is still locked in the radioactive core of the reactor. This accident caused great changes in safety instrumentation, control room design, manpower training, and Nuclear Regulatory Commission rules and procedures.

With 81 stations in operation in the U.S. in 1981, hope is expressed for a resumption of nuclear plant design, but the high capital costs seem to preclude that possibility in the U.S. even if environmental and safety questions be resolved. Such an unfavorable view of nuclear energy is not prevalent in Europe.

*The control panel at a nuclear-powered generating plant.*

# ELECTRIC POWER RESEARCH INSTITUTE

Prior to the 1960's, the utilities' technical and research needs had been met by the equipment manufacturers, most of which were established at the birth of the central station industry and had grown with it. The research and development carried on was therefore largely equipment oriented, with system design and planning relegated to minor positions.

As the utility systems became larger and more interconnected, performance and system reliability remained the responsibility of the individual operating utilities, except in a few very large systems. Most of the operating companies were local or regional public service monopolies, each independently managed and under state and federal regulation. Because rates were controlled to minimize consumer costs, it was difficult for even the largest utilities to justify the costs associated with new development or research work.

Growing awareness that the manufacturers and utilities were incapable of solving these abstract, system-oriented problems led in 1979 to the formation of Electric Power Research Institute (EPRI), a nonprofit organization to manage a national research and development program for the electric power industry. This institute now selects and funds research projects to provide answers for present and future electric energy needs.

Support of EPRI is voluntary, but of 3000 electric utilities composing the U.S. industry, 571 are voluntary members, representing about 70 percent of the total generation capacity in the U.S. Annual assessments are based on kilowatt hours sold, and in 1982 amounted to about $300 million. Research is contracted with industry, university, and government laboratories, to take advantage of all available professional staff and facilities.

The program, managed largely through advisory committees, has developed with the needs created by the oil crisis, the requirements for conservation of fuel, and the increasing cost of new capital, which makes it economical to extend the life of older plants and equipment. Other technical problems include the cleaning up of undesirable effluents, the higher temperatures and pressures met in today's sophisticated generation equipment, and system interconnections.

This cooperative effort has made possible much operational and system research not possible in the pre-war era of dependence on vendor research. However, the vendors have continued to supply new equipment and innovations as industry needs have developed.

## For Further Reading

R. Belfield, "The Niagara system: The evolution of an electric power complex at Niagara Falls, 1883–1896," *Proc. IEEE,* vol. 64, p. 1344, Sept. 1976.

C. A. Berg, "Process innovation and changes in industrial energy use," in *Energy II.* Washington, DC: Amer. Ass. Advance. Sci., p. 3, 1978.

H. Dorn, "Hugh Lincoln Cooper and the first detente," *Tech. Cult.,* vol. 20, p. 322, Apr. 1979.

L. Hannah, *Electricity Before Nationalisation.* Baltimore, MD: Johns Hopkins Univ. Press, 1979.

T. J. Healy, *Energy, Electrical Power, and Men.* San Francisco, CA: Boyd and Fraser, 1974.

T. P. Hughes, "The Electrification of America: The system builders," *Tech. Cult.,* vol. 20, p. 124, 1979.

——, *Networks of Power.* Baltimore, MD: Johns Hopkins Univ. Press, 1983.

——, "Technology and public policy: The failure of giant power," *Proc. IEEE,* vol. 64, p. 1361, Sept. 1976.

T. K. McCraw, "Triumph and irony — The TVA," *Proc. IEEE,* vol. 64, p. 1372, 1976.

Special Issue on Three Mile Island and the Future of Nuclear Power, *IEEE Spectrum,* vol. 16, Nov. 1979.

P. Sporn, *Technology, Engineering, and Economics.* Cambridge, MA: M.I.T. Press, 1969.

# 7

# ELECTRONS AND HOLES: MUCH FROM LITTLE

As radio began its expansion in 1922, the annual sales of industries associated with radio and vacuum tubes in the U.S. totaled a mere $60 million. By 1960, the computer, television, and other new industries had raised that figure to $13 billion. By the end of the first century of the electronic age in 1983, total sales were nearly $150 billion from those industries in the U.S. The worldwide figure for 1983 was estimated to be approaching $500 billion dollars. That stunning figure would have been reduced by a factor of a hundred or more if the industry had not embraced the transistor and other solid-state inventions made since 1947. This chapter, then, recounts the coming of the solid-state age.

## A NAME FOR
## A NEW FIELD

By 1929, the vacuum tube had become known as the "electron tube," but there was no generic name for the new industries based on electron tubes. The omission was corrected in that year by O. H. Caldwell and Keith Henney, who were planning a new magazine to be published by McGraw-Hill. According to Caldwell, who was searching for a name for the venture, the term "electronics" was suggested to him by

Zay Jeffries of G.E. Henney remembers, on the other hand, that the name was suggested by John Mills of Bell Telephone Laboratories. The adjective "electronic" had appeared in a British journal and elsewhere, but the noun form, for the profession and for the industry, was new.

The first issue of *Electronics* appeared in April 1930, and the new word caught on quickly. In 1932 Prof. Truman Gray of M.I.T. was inspired by the magazine to name a course "Engineering Electronics." One of the first books to use the new word in its title was published in 1938, written by a young staff member of *Electronics* who had been a student in the first year of Prof. Gray's course. The title of the book was, of course, *Engineering Electronics*.

During the first half century of the electronics age, the technology was dominated by the electron tube. This device appeared in so many forms, over such a range of power levels, and in so many different applications that even to list them would be a major exercise. It is sufficient to repeat the generally accepted verdict of military historians: the Allies' victory in World War II would not have been possible without their greatly superior electronic systems. Many examples of such systems come to mind: radar, aircraft communications, identification of friend from foe, the vacuum tube artillery fuze, and electronic countermeasures and countercountermeasures (see Chapter 8). All of these developments and more like them were based solidly on the electron tube.

Actually, only one other electronic device was available. That was the solid-state diode rectifier. It allowed one-way-only flow of current, as did Edison's lamp with a plate. But it differed from all other electronic devices of the time because the current in it proceeded through a solid, a type of solid known as a semiconductor. The solid-state diode was brought to a high standard of performance during the war years. It was, in fact, the cornerstone on which the second half of electronics' first century was built.

# SEMICONDUCTORS: KEY TO SOLID-STATE ELECTRONICS

This chapter is confined to the period 1940–1980, during which the solid-state electronics revolution took place. This concentration on recent history is justified by its impact on the world. In those four decades, the appearance of solid-state devices forced the abandonment of nearly all electron tubes and their works.

Today, there remain only three fields in which vacuum electronics still serves: 1) that where the transparency of glass is essential, as in television tubes, computer displays, and some photoelectric devices; 2) the generation of large amounts of power at very high frequencies; and 3) the successors of the first applied electronic device, the 1895 X-ray tube of Wilhelm Roentgen.

The solid-state diode rectifier was the first widely used semiconductor device. As the name suggests, a semiconductor has properties intermediate to those of a conductor like copper and an insulator like glass. The electrons in copper are free to move easily, and thus act as a current. The electrons in glass, however, are so tightly bound to their atoms that they cannot readily move.

*Wilhelm K. Roentgen, the young German Professor of Physics whose discovery of invisible radiation ushered in a new era in physics and medicine.*

# DEVELOPMENTS DURING WORLD WAR II

Early in 1941, during the war in Europe, the M.I.T. Radiation Laboratory was founded to work on the development of microwave radar and other war-related projects (see Chapter 8). To it were recruited the cream of the academic physics community. Attention focused on the semiconductor diode as the detector of weak radar signals returning from their targets. The solid-state diode was preferred over its vacuum-tube cousin, the diode tube, because it had far less capacitance and was thus a more efficient detector of high-frequency radar signals. The Radiation Laboratory had the primary responsibility for research directed toward improvement of such diodes, but it farmed out some of this work to teams at academic physics departments.

One such team, at Purdue University, consisted of gifted graduate students led by Dr. Karl Lark-Horovitz. Lark-Horovitz decided initially to concentrate on a particular semiconductor, germanium. He favored germanium because it was an element and could be refined to a high state of purity, unlike other semiconductors that were compounds of such elements as silver, lead, copper, sulfur, and oxygen.

At that time, the elements silicon and germanium were thought to possess semiconducting properties only in the impure state. When Lark-Horovitz produced pure polycrystals, he reported to his sponsors at M.I.T. that extremely pure germanium and silicon were semiconductors also, and developed properties predictable from their impurity content.

His team produced their first germanium diodes in 1942. Their uniformly high performance and stability under difficult conditions provided early evidence of a powerful truth: great care in the preparation of pure semiconductors would yield large dividends. It was clear that the polycrystalline forms of germanium and silicon would not suffice. The crystal lattice had to be so regular and unbroken that only perfect single crystals would serve. No substance in its natural or refined state had ever met this requirement. To achieve the required purity and regularity of structure required great improvements in metallurgical techniques.

Lark-Horovitz and his team continued their work well after the end of the war, and they missed inventing the transistor by only a few weeks. That near-miss can perhaps be explained by noting that Lark-Horovitz and his students were engaged in academic research and interested in the discovery of physical truth. In contrast, at the Bell Laboratories, where the invention did occur two days before Christmas in 1947, the search for a device to replace the vacuum tube had been in progress since before the war.

# WHY TRANSISTORS?

The vacuum tube—for all its ability to oscillate, modulate, demodulate, and amplify at any frequency from zero (direct current) to many millions of cycles per second—was an ornery beast. It was bulky, it used too much power for heating its filament or cathode, and it wasted power in heating its plate. It required voltages in the hundreds or thousands. Above all, it could not be relied upon to operate for long periods of time to

*From left: John Bardeen, William Shockley, and Walter Brattain—Nobel Prize winners in 1956 for their invention of the transistor.*

match the virtually indefinite life of the other components in electronic systems. Sooner or later, its emission of electrons would be reduced and it would have to be replaced. The replacement cost and the uncertainty of the timing of the failure caused great grief in the repeater amplifiers, the carrier modulators, and the undersea cables of the American Telephone and Telegraph Company. So it was not surprising that the Bell Laboratories were searching for something better.

Another problem faced by AT&T was the growth in telephone traffic through its switching centers, whose electromechanical relays were bulky and slow. In 1936, Dr. Mervin Kelly, then Director of Research of Bell Laboratories, suggested that eventually those relays must be replaced by faster and longer-lived devices. His vision was not fulfilled for years, but it remained as a guiding light for three people whose place in history is now assured, for they conceived what is clearly the greatest invention of the 20th century—the transistor.

## THE THREE INVENTORS

In 1936, Dr. William Shockley joined the staff of the Bell Laboratories. His primary interest was the behavior of electrons in solids. During the war, he served as an expert consultant to Secretary of War Stimson, and on his return he was given joint charge, with Stanley Morgan, of Bell's research in solid-state physics.

Also in 1945, the Labs recruited another physicist to the team. This was Dr. John Bardeen, a theoretical physicist of first rank, the only man to win two Nobel Prizes in physics: one for the transistor work at Bell and the other for later work in superconductivity at the University of Illinois. The third member of the transistor triumvirate was Dr. Walter Brattain, a gifted experimental physicist who had joined the Bell Laboratories staff in 1929.

In 1956, these three men were awarded the Nobel Prize in physics for the invention of the transistor. Actually, two different types of transistor were covered by the award: the point-contact transistor invented by Bardeen and Brattain in 1947, and the junction transistor conceived by Shockley in 1948 and reduced to practice in 1951.

The point-contact transistor opened the door, but it was difficult to manufacture and limited in performance. It was soon supplanted by generations of transistors and integrated circuits based on the junction transistor principle.

Still another transistor principle had occurred to the Shockley team several years before the concept of the point-contact transistor. This was the field-effect transistor, but it resisted all attempts to reduce it to practice. This had been the fate of similar work by J. E. Lilienfeld, a German emigré to the U.S., when in 1925–1928 he patented what appears to have been a field-effect transistor, but which could not then be produced. Shockley's field-effect transistor was finally realized in 1954, as a result of improved materials and processes, and is now a central feature of solid-state electronics.

That the Bell Laboratories demonstrated the group-research philosophy when it assembled a multidisciplinary team to attack the semiconduuctor problem is well known. The advantages of such an attack were illustrated by Brattain, who tells of the spirit of mutual learning: "We went to the heart of many things during the existence of this group and always when we got to the place where something had to be done, experimental or theoretical, there was never any question as to who was the appropriate man in the group to do it."

# THE MINORITY
# MOVEMENT IN
# SEMICONDUCTORS

At very low temperatures, pure semiconductors do not conduct electricity, as all of the electrons are bound to the atoms of the crystal. As the temperature rises, the vibration of the atoms produced by their heat energy causes some of the electrons to shake free. These electrons can then move when a voltage is applied, appearing as a current of negative charge. But the vacant electron sites (holes) are, in effect, positive charges, and these vacancies can also move, in much the same way that the gap between dominoes lined up in a row "moves" when they fall one after the other. This movement of vacancies constitutes a current of positive charges, opposite in direction to the electron movement.

Now consider germanium to which an infinitesimal amount of phosphorus has been added. The phosphorus atoms add an extra electron which can move freely from atom to atom within the crystal lattice. This type of germanium, man-made with an extra supply of negative electrons, is called "n-type" germanium. The electrons, in copious supply, are called the "majority carriers" of the electric current. Also present are the holes created when electrons are broken free by thermal energy. These holes are much less numerous than the impurity electrons and are known as "minority carriers."

If the impurity added to the germanium is boron, the roles of the electrons and holes are reversed. Boron introduced into the germanium lattice has a deficiency of electrons; that is, it naturally creates holes that are free to move in the lattice. These positive holes are the majority carriers in "p-type" germanium. The thermally generated electrons, in smaller numbers, are the minority carriers in the p-type material. It is the minority carriers that are controlled in transistors.

# THE POINT-CONTACT
# TRANSISTOR IS
# INVENTED

Single crystals of germanium are grown from a pot of melted element into which is dipped a small single crystal of germanium. As this is rotated and slowly withdrawn, germanium freezes around it and fits itself to the arrangement of atoms in the seed crystal. This crystal growing process was developed by Czochralski and perfected by Teal and Little of the Bell Laboratories metallurgical staff. Teal later left Bell to join Texas Instruments Corporation, where he successfully adapted the technique to the growing of single crystals of silicon, now used in most semiconductor devices.

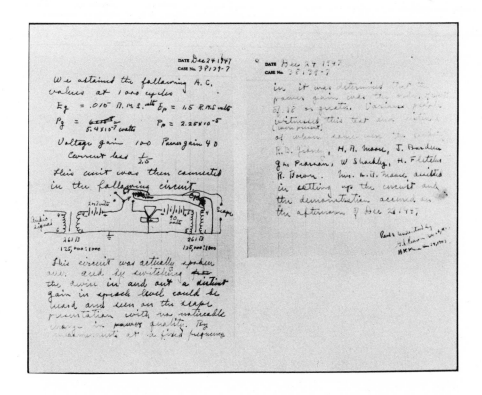

*A page from the laboratory notebook reporting the first successful experiments with transistor action.*

To introduce the desired impurity, the phosphorus or boron was added to the pot in measured amounts. Later, much more sophisticated techniques were adopted, from alloying and diffusion to ion implantation. But solid-state techniques would have come to naught had it not been for the metallurgical techniques of Czochralski and Teal and his colleagues.

Single crystals of n-type and p-type germanium were available to Bardeen and Brattain in 1947. Using n-material, they found that if a pointed wire was pushed against the crystal, with a battery making the wire positive and the crystal negative, the positive point would attract the majority carrier electrons, and a strong current was produced. But if the battery terminals were reversed, the negatively charged wire would repel the electrons and attract the much smaller number of positive holes.

Bardeen and Brattain decided that control of the minority current might be possible by placing a second pointed wire very close to the first on the crystal surface. To their delight they found, on December 23, 1947, that a current could be drawn from the n-germanium by the second wire, and that this current was an amplified copy of any changes in the current in the first wire. The transistor had been invented. Amplification had occurred in a semiconducting solid-state device.

Bardeen, the theoretician, explained what was going on within the germanium crystal. He was the ideal teammate for Brattain, the experimentalist who "felt in his bones" what to do next. There followed months of strictest secrecy while patent applications were filed and the strategy of public announcement was arranged. By June 1948, the news was out.

The entrenched body of vacuum-tube engineers found little to admire in the

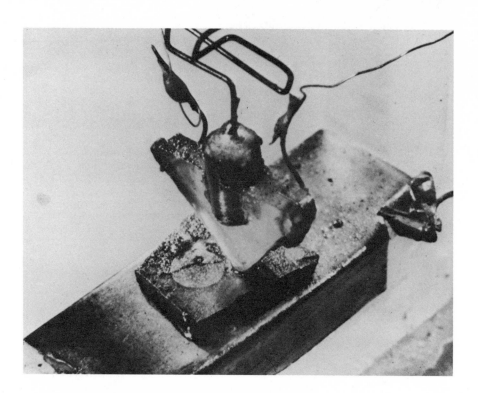

*The first point-contact transistor,*
*a far cry from the infinitesimal*
*microcomponents of the 1980's.*

point-contact transistor and little fear in it as a competitor. They were wrong, but it took years for their culture to fade away.

The invention of the transistor exemplifies those quirks of timing that determine whose names go down in history. The work on germanium and silicon at Purdue University by the students of Lark-Horovitz was highly appreciated by the scientific community after wartime restrictions on publication were lifted. In early 1948, two members of the Purdue team, Ralph Bray and Seymour Benzer, reported at a meeting of the American Physical Society on experiments with point contacts on germanium. Brattain, who was in the audience, knew that the effects described were due to minority carriers and realized how close Bray and Benzer had come to inventing the transistor. Bray later said that if his pointed wire had been closer to Benzer's electrode, they might have observed transistor action.

# THE P-N JUNCTION: KEY
# TO THE JUNCTION
# TRANSISTOR

A junction is formed in a semiconductor crystal by change of impurity, from n to p. The change is not structural, and the two regions are part of the same crystal. Let us look at the n side of the junction, where free electrons are majority carriers and a few free holes are the minority carriers. The reverse situation is present on the p side of the junction. All these charges have random and erratic motions due to their heat energies. Left to their own interactions, some of the majority electrons in the n region drift or "diffuse" to the p side, where they encounter a supply of holes. Holes, being electron vacancies in the atomic structure, gobble up and immobilize electrons that they meet. Similarly, some majority holes in the p side diffuse to the n side, and gobble up and immobilize electrons there. These two acts of cannibalism or "recombination" cause a shortage of mobile holes and mobile electrons on either side of the junction region, producing a depletion layer in which there are no mobile charges, and this serves as a barrier to charge motion or current through the layer.

If we connect a battery across the junction, positive terminal to the n side, and negative terminal to the p side, this draws more electrons away from the junction on the p side, or the depletion layer is thickened. This is the current-off connection of the battery.

When the battery connections are reversed, that is, with the positive terminal to the p germanium, negative electrons are taken from their parent atoms and move into the battery. The holes so created are filled by other electrons from deeper in the material, and a current of electrons passes through the p material. For the current to continue, electrons are attracted across the boundary layer from the electron-rich n side. Thus, the depletion layer is narrowed and electrons are pushed across. This is the current-on connection of the battery, positive to the p side. Such a junction carries current with one polarity of the battery and does not carry current with the reverse polarity.

# THE JUNCTION
# TRANSISTOR

The n-p-n (or p-n-p) junction transistor has a second closely associated junction with the intermediate p (or n) region made very thin. These three regions are named the emitter, the base, and the collector.

In the so-called common-base connection of an n-p-n transistor, the input current is injected into the n-type emitter. With the battery negative to the emitter and positive to the collector, the injected electrons move readily across the first junction and into the inner p base. On entering the base, the electrons become minority carriers. This minority electron current is controlled by the transistor. A few of the electrons fall prey to the holes in the p base,

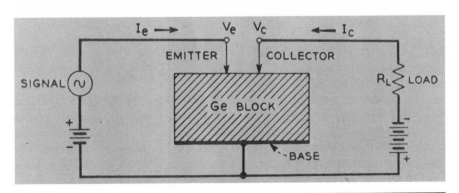

A schematic diagram of a simple transistor amplifying circuit (from an early paper by Bardeen and Brattain). $I_e$ and $V_e$ are positive (current and voltage, respectively); $I_c$ and $V_c$ are negative.

but since the base is very thin most of the electrons survive and drift to the second p-n junction.

The positive battery terminal is connected to the n collector. Holes in the inner base thus face a repelling force at the second (base-collector) junction, but the junction layer is transparent to the electrons moving out of the base to the positive collector. On passing into the collector these electrons are attracted to the positive battery terminal, which collects them.

The amplification occurs at the second junction, between base and collector. In passing this junction, the electrons have their energy increased, this energy coming from the collector-base potential. Electrons are injected into the emitter at a low potential, possibly a few tenths of a volt; they are extracted from the collector by a higher potential of several volts. Obviously, the power level has been increased; this is amplification.

## ACHIEVEMENTS OF THE JUNCTION TRANSISTOR

To achieve transistor action and gain, several details have to be properly adjusted. The tiny fraction of impurities in the n and p materials, roughly one impurity atom per $10^{10}$ semiconductor atoms, has to be precisely maintained in raw material production. Otherwise losses of electrons and hole by recombination will reach too high a level. The base must be very thin so that the time for charge drift across the base is very short. Otherwise, the transistor cannot respond to rapid changes in its input current.

The transistor far surpasses the vacuum tube. It is small, and can be made so small in integrated circuits that it cannot be seen by the unaided eye. It has no heated filament and operates instantly upon the application of its voltages, usually low in value. It is efficient, making it ideal for portable, battery-powered equipment.

Most important, it has a long, essentially indefinite, life. The transfer of electrons or holes is not a destructive process, as is the thermal emission of electrons in vacuum tubes. It was the complete answer to the telephone companies' prayer for something better than the tube, and it has led to generations of new electronic systems.

## CHANGING FORMS OF THE JUNCTION TRANSISTOR

Shockley's junction design came in 1951, and many improved forms of junction transistor followed. In 1953, David Smith and William Bradley of the Philco Corporation offered the surface-barrier transistor. This used a base of germanium, etched to extreme thinness by chemical action, with metals alloyed on either side as emitter and collector. The thin base made it useful for high frequency action. Then came the micro-alloy form in 1954 and a little later the diffused-base germanium transistor.

*Jack S. Kilby holding his early integrated circuit.*

In 1954 appeared the first commercial silicon transistors from the Texas Instruments Corporation. Silicon is the second most plentiful element in the earth's crust — it is a constituent of beach sand and rock, but it is inferior to germanium in the speed of transit of minority carriers. Silicon dioxide, easily manufactured and worked, is an excellent and stable insulator for integrated circuits. Silicon transistors are operable at higher temperatures than are those of germanium. At present, silicon is the material of choice for nearly all solid-state circuit devices, the only exception being a new compound semiconductor, gallium arsenide, used in devices operating at hundreds of billions of cycles per second.

In 1958 came the mesa transistor, named for the shape of the diffused and etched island of silicon, within which the n-p-n or p-n-p structure is formed. Exploited heavily by Motorola and Texas Instruments, this transistor made the Philco transistors obsolete only five years after they were first introduced. In 1960 came the planar transistor from John Hoerni at Fairchild Instrument, in which a flat plate of silicon was exposed at high temperatures to successive diffusions of impurity atoms supplied in gaseous form above its surface. The planar process remains important for making discrete transistors, but its most important contribution to the art was that it led to the solid-state integrated circuit chip.

# TRANSISTORS,
# ET CETERA, ON A CHIP

During the 1950's, the assembly of electronic equipment was a time-consuming process, slowed by the complexity of the circuits; typical computers, for example, were using 100 000 diodes and 25 000 transistors. The enormous growth of solid-state electronics did not start, indeed could not start, until a way was found to meet these complex needs with simplified components and to compress thousands of devices into a small space with simple and reliable connections to the outside world.

The concept of putting several components together in integrated form appeared in 1953, when Harwick Johnson of RCA applied for a patent in which transistors, resistors, and capacitors were combined in a single piece of semiconducting material. Sidney Darlington of Bell Laboratories and Bernard Oliver of Hewlett-Packard made similar proposals but did not include the use of passive resistors and capacitors.

The year of the chip was 1959. In February, Jack Kilby of Texas Instruments applied for a patent describing two circuits in which were included junction transistors with resistors and capacitors formed within a chip of germanium, the parts interconnected by gold wires external to the chip. In July, Robert Noyce of Fairchild Instruments applied for a patent in which he described planar elements on a chip of germanium interconnected by deposited aluminum strips. Noyce's patent contained all the methods used today in the production of integrated circuits. The others named above made major contributions, but it was Noyce who brought it all together in a workable system.

Noyce's career well illustrates the upward mobility of scientists who possess strong managerial talent. This has produced a number of leading engineer—managers in the industry today. In Noyce's case, he joined the research staff of Philco Corporation in 1953 with a Ph.D. in physics from M.I.T. and worked on the surface-barrier and micro-alloy transistors. Offered a job by Shockley, who had set up his own company in Palo Alto, CA, Noyce stayed with Shockley for two years. He then transferred to the Fairchild Camera and Instrument Corporation and did work that led to his planar integrated circuit patent.

Asked to transfer to fill in temporarily for an absent manager, Noyce quickly

*Andrew Grove, Robert Noyce, and Gordon Moore, the founders of Intel Corporation.*

**1958**

← - - - - 7/16 in. - - - - →

*This unassuming device is actually the first Texas Instruments integrated circuit.*

learned that managing can be as exciting as physics. In 1967, he left Fairchild and, with Gordon Moore, founded the Intel Corporation. There, a few years later, the microprocessor for elementary computing was invented. Needless to say, today Robert Noyce is rich. In 1978, the IEEE awarded him its highest accolade, the Medal of Honor. Two others similarly honored are John Bardeen (1971) and William Shockley (1980).

## HOW INTEGRATED CIRCUITS ARE MADE

The secret of the integrated circuit is that the same materials and operations which in 1960 produced one transistor today produce hundreds of thousands. The number of active elements — transistors and diodes — packed onto a 6-mm (0.25-in) square chip of silicon has increased a hundredfold every decade since 1960.

The integrated circuit consists of numerous interconnected layers of silicon with differing impurity contents. The pattern for each layer's configuration is drawn to scale, then optically reduced to form a set of masks. A great many circuit masks appear side by side on a master, which can cover a 4 in diameter silicon wafer. This thin wafer is coated with a photoresist chemical that responds to light reaching it through the mask, just as in making an engraving for a newspaper. After exposure, the unexposed photoresist is washed away, leaving the silicon wafer partially uncovered and partially protected. The exposed parts of the silicon are processed by gaseous diffusion of impurity elements to form the elements of transistors or diodes. Insulation between layers is provided by oxidizing exposed silicon areas to silicon dioxide, and interconnections are made by depositing very narrow strips of aluminum.

*Left: A 1961 Fairchild integrated circuit. Compare this to the microcircuits pictured in Chapter 10. Right: Here a worker at RCA is triggering a laser to trim the excess material from a tiny circuit chip. She monitors the process on the television screen.*

After completion of the array of circuits, the wafer is broken apart to produce individual tiny chips. A single grain of dust can ruin a chip, so high standards of cleanliness and dust-free procedures are enforced in the production process. After the chips pass final checkout, they are encapsulated in a flat plastic package, along the edges of which are metal pins connected to the chip circuits inside.

# THE MANY USES
# OF SOLID-STATE
# ELECTRONICS

Solid-state devices are made primarily for operation with on–off (digital) current pulses as in computers, but are also used for continuously changing (analog) signals as in TV sets, radios, and circuits to process and control large amounts of electric power. They can also generate or be responsive to light and other radiation.

Pulse operation in computers and information processing is covered in Chapter 10. In radios, the transistors amplify or produce currents which are enlarged copies of the input current changes. In a phonograph system, the transistors may take a few millionths of a watt generated by the phonograph cartridge and produce hundreds of watts of output power in the loudspeakers. Today's television sets provide a complex performance for a modest cost — less than $75 for a black-and-white receiver. The signals are handled by transistors much as in a radio receiver, although many auxiliary functions are now generated by circuits on chips.

Power electronics arose from development of a practical silicon controlled rectifier (SCR) by R. A. York of the General Electric Company in 1956. These devices handle heavy currents, which can be turned on but not off — the latter operation occurs only when the external power is shut off. In ac systems, this happens every cycle, so the device is practical on ac with conduction over controlled parts of each cycle. The average value of the current is controlled. This task was formerly performed by now-obsolete mercury vapor tubes.

At 25 or 60 Hz, the SCR need not have a small base region as capacitance is not of consequence; therefore, SCR's can be made large enough to carry thousands of amperes. Arrays of SCR's are used at large power centers to convert ac to dc for long-distance transmission. At the other end of the line, a similar array performs the inverse operation, producing alternating current of desired frequency for local needs.

There are light-sensitive and light-emitting diodes, both used in information displays. The solid-state laser has more pervasive applications. It produces light in a very narrow and homogeneous band of wavelengths. A major application is optical fiber communications, where the infrared radiation travels down a mile or more of thin glass fiber while carrying a telephone voice or data message. At the end, a solid-state diode transforms the message-carrying laser light back to electrical form for retransmission or transfer to the existing wire telephone network.

Many more voice channels can be carried in one glass fiber than can be carried with one electric wire. By 1983, the technical feasibility of such a system had been proven in the heavy-traffic corridor between Boston and Washington. Another common application is at the supermarket checkout counter, in the lasers and diodes used to read the bar codes containing pricing information on the product packages.

Every reader of this book is surrounded by examples of the solid-state electronics revolution. What next commands our attention is the application of electronic devices in the making of war, and the synergy that exists between their military and civilian uses.

## For Further Reading

Braun and McDonald, *Revolutions in Miniature: The History and Impact of Semiconductor Electronics*. Cambridge, England: Cambridge Univ. Press, 1978.

C. Hogan, "Reflections on the past and thoughts about the future of semiconductor technology," *Interface Age,* Mar. 1977.

M. E. Jones, W. C. Holton, and R. Stratton, "Semiconductors: The key to computational plenty," *Proc. IEEE,* vol. 70, pp. 1380–1409, Dec. 1982.

C. Susskind, "The origin of the word 'electronics,'" *IEEE Spectrum,* vol. 3, pp. 72–79, May 1966.

C. Weiner, "How the transistor emerged," *IEEE Spectrum,* p. 24, Jan. 1973.

M. F. Wolff, "The genesis of the integrated circuit," *IEEE Spectrum,* p. 44, Aug. 1976.

——, "The R&D 'bootleggers': Inventing against odds," *IEEE Spectrum,* p. 38, July 1975.

# 8

# ELECTRONS IN WAR AND PEACE

Throughout recorded history, the technical arts have provided the weapons of war and the defenses against them. Military engineering was once the only organized branch of technology, and was practiced by such men as Leonardo da Vinci and Michelangelo. When engineers first organized in support of technology for the civilian world in 1818, they called themselves civil engineers to distinguish their work from that of their military counterparts. Preparing for war, waging it, defending against it, and finding means to prevent it have been major occupations of electrical engineers throughout the IEEE's history.

## WAR STIMULATES INVENTION

Major Edwin H. Armstrong's great invention, the superheterodyne circuit, basic to billions of radio and television receivers, was conceived as a more sensitive means of reception when he was a Signal Corps officer in Paris in World War I. The first electronic computers, COLOSSUS and ENIAC, were built during World War II, one to break enemy codes and the other to compute artillery ballistics tables. Three great men of information theory, Shannon, Weiner, and von Neumann, achieved new and major insights while working under wartime pressures. During the war, the work on semiconductors at Purdue University by Lark-Horovitz and his students contributed greatly to the development of solid-state electronics.

We noted in the last chapter the judgment of military historians concerning the importance of superior electronic systems to the Allies' World War II victory. This lesson has not been lost on those charged with military preparedness. The U.S. Defense Department's current outlay for research and design work, and the procurement of electronic weapons and computer systems, is estimated to be nearly one half of its budget.

This chapter describes systems devised by electronics engineers and scientists for military purposes, but is limited to those that have important uses in times of peace. The projects chosen for review illustrate the synergy between military and civilian engineering: radar, sonar, electronic aids to navigation, and mobile communications. Others, such as information theory, the computer, and the metallurgy of the solid state are treated elsewhere in this book. The proximity fuze, guided weapons, and other developments of less importance to civilian life have been omitted. Added is one use that arose after World War II: satellite communications and observation of the earth. The latter technology is of great importance in keeping peace amid threats of nuclear war.

# RADAR BEGAN
# WITH HERTZ

The word "radar" was invented in 1941 by the U.S. Navy as an abbreviation of "radio detection and ranging." Radio waves are reflected by objects whose electrical characteristics differ from those of their surroundings, and the reflected waves can be detected. The electrical

*Heinrich Rudolf Hertz.*

*Left: Oliver Heaviside, the eccentric British telegrapher who tamed Maxwell's mathematics and predicted the existence of the ionosphere.
Right: Arthur Kennelly, codiscoverer with Heaviside of the ionosphere. He worked for a time as an assistant to Thomas Edison — who was baffled by mathematics — and in 1902 became a Professor at Harvard.*

differences may be very subtle — the radar carried by commercial airliners can distinguish between calm and turbulent air, permitting the pilot to avoid uncomfortable or dangerous passages.

Radar is based on much early research. In 1887, Heinrich Hertz proved that radio waves exist and can be reflected. He set up laboratory apparatus to test Maxwell's proposal that there were various forms of radiant electromagnetic energy, of which light was only one. Hertz passed a spark across a gap and found that a weak spark simultaneously jumped across a small gap in a copper ring at the opposite end of the room. He discovered that he could block the effect by placing a metal plate between the gaps and that a plate placed to the side of the blocked path would reflect the waves. The waves Hertz used varied in length from 66 cm to several meters. More than half a century was to pass before reflections of similar waves were used in the radar of World War II.

In 1902 Kennelly and Heaviside suggested that the upper levels of the atmosphere were ionized, that is, possessed a high concentration of electrically charged particles. Transmission of radio signals beyond the horizon, as in the transatlantic tests of Marconi in 1901, could be explained as reflections from such an ionized layer. In 1924, Appleton and Barnett of England proved the existence of this layer by reflecting radio waves from it. They used waves of varying frequency, so that by the time the reflected wave returned to the ground, the transmitted frequency had changed. The difference in frequency of the two signals could be interpreted as a time difference. Using the known velocity of radio waves (about $3 \times 10^8$ m/s or 186 000 mi/s), the height of the reflecting layer could be computed.

In 1925, Breit and Tuve of the Carnegie Institution in Washington refined this method by transmitting short bursts (pulses) of radio energy toward the sky, and measuring the time required for the reflected signals to return to earth. This technique showed that the ionosphere, as it was called, was composed of several layers whose ionization varied with the time of day and the season. Such information is now used to choose the optimum wavelength for short-wave communication.

A more secret development of radar began in 1922, when A. Hoyt Taylor and

L. C. Young of the U.S. Naval Research Laboratory in Anacostia reported that the transmission of 5-m radio waves was affected by the movements of naval vessels on the nearby Anacostia river; they also suggested that the movements of ships could be detected by such variations in the signals. Returning to this work in 1930, Dr. Taylor succeeded in designing equipment to detect moving aircraft with reflected 5-m waves.

In 1932, the Secretary of the Navy communicated Taylor's findings to the Secretary of War to allow application to the Army's antiaircraft weapons; the secret work continued at the Signal Corps Laboratories. In 1934, a 0.5-W source of 10-cm waves was used to confirm Taylor's work. In July of 1934, the work of Breit and Tuve was recalled by the Director of the Signal Corps Laboratories, Major Blair, who proposed "a scheme of projecting an interrupted sequence of trains of oscillations against the target and attempting to detect the echoes during the interstices between the projections." This was the first proposal in the U.S. for a pulsed-wave radar.

By 1936, the Signal Corps had begun work on the first Army antiaircraft radar, the SCR 268, and a prototype was demonstrated to the Secretary of War in May 1937. It operated at a wavelength of 3 m, later reduced to 1.5 m. Navy designs for shipboard use operated at 50 cm and later 20 cm, the shorter wavelengths improving the accuracy of the direction-finding operation.

Earlier, Sir Robert Watson-Watt of the National Physical Laboratories in England had independently worked along the same lines, and he successfully tracked aircraft in 1935. The urgency felt in Britain was such that they also started work on "early warning" radar using 12-m pulsed waves. The first of these "Chain Home" stations was put in place along the Thames estuary in 1937, and when Germany occupied Czechoslovakia in November 1938, these stations and others were put on continuous watch and remained so for the duration of the ensuing war. In 1940, British ground-based radars were used to control 700 British fighter aircraft defending against 2000 invading

German planes. This radar control of fighter planes has been credited with the decisive role in the German defeat in the battle of Britain.

# MICROWAVE RADAR

Antenna dimensions must be large compared with the radio wavelength to obtain a narrow beam of energy. The wavelengths of the early radars required antennas too large for mounting in aircraft, yet the need was apparent. If a fighter pilot could detect an enemy aircraft before it came into view, he could engage the enemy with advantages of timing and maneuver not otherwise obtainable. On bombing missions, airborne radar could assist in navigating to and locating targets.

The prime need was for a high-power transmitting tube, capable of pulse outputs of 100 000 W or more at wavelengths of 10 cm or less. Such powers were needed because the tiny signal reflected from a distant target back to the radar receiver was otherwise undetectable.

An answer to the problem, found by the British in 1940, was a new form of vacuum tube, the cavity magnetron. In it, dense clouds of electrons swirl in a magnetic field past cavities cut in a copper ring. The electrical effect is much like the acoustic effect of blowing across the mouth of a bottle. The electrical oscillations built up across the cavities produce powerful waves of very short wavelength. It was invented at the University of Birmingham; the British General Electric Company produced a production design in July 1940.

The scientific advisers to Prime Minister Winston Churchill faced a problem — active development of microwave radar would create critical shortages of men and material then devoted to extending the long-range

*A V-beam radar installation.*

*Above: Karl Taylor Compton.*
*Right: The assembly of an SWT cavity magnetron at the M.I.T. Radiation Laboratory in November 1944.*

radar defenses. They suggested to Churchill that the Americans, not yet at war, be shown the new magnetron and asked to use it to develop microwave radar systems.

The Americans were then forming an organization to handle this problem. Dr. Vannevar Bush had requested authority from the President in June 1940, to set up the National Defense Research Committee. Granted this, he established five NDRC divisions, one to be concerned with "detection" with M.I.T. President Karl T. Compton in charge. Alfred Loomis, a financier with a flair for science, was to head the Microwave Committee. In August 1940, that committee met with British representatives who had with them a sample of the cavity magnetron.

The American response was enthusiastic and in September they set up a new center for microwave radar research and development, the M.I.T. Radiation Laboratory. Dr. Lee Dubridge, then head of the Department of Physics of the University of Rochester, was appointed its director. Concurrently, the British magnetron was copied at the Bell Laboratories and judged ready to be put into production.

At that time the technology of microwaves was in its infancy. Little had been translated from basic equations to hardware and there were few engineers engaged in the field. Those who were available were asked to join the M.I.T. "Rad Lab," but the main source of talent was the community of research-oriented academic physicists. Among those recruited to the Laboratory were three men who later were awarded Nobel Prizes in physics: I. I. Rabi, E. M. Purcell, and L. W. Alvarez.

The men recruited from the academic laboratories were able to shift quickly from theory to the building of prototype equipment. The leaders also mastered

 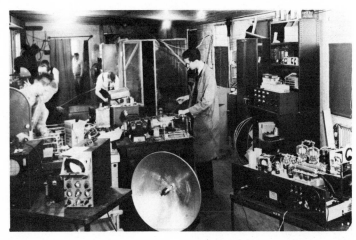

the art of establishing liaisons with both the military and engineers in industry. Never has the close relation between physics and electronics engineering been better demonstrated.

By 1942, a cavity magnetron capable of producing a peak pulse power of 2 million W was available. The model SCR 584 10-cm ground radar for tracking enemy aircraft and leading friendly ones home was an early success. It was designed by a team led by Dr. Ivan Getting (IEEE President, 1978). Smaller and lighter radars operating at 3 cm and 1 cm were quickly adopted by bomber missions for location of ground targets, and ground-based radars for early warning of impending attack were developed.

*Left: An aerial view of the M.I.T. campus in 1945, looking west. The Radiation Laboratory buildings are at right center.*
*Right: The Roof Laboratory main room at the Radiation Laboratory.*

# COUNTERMEASURES

Since radar echoes are very weak, they are easily obliterated by strong signals sent out on the same wavelength. Such intentional interference (jamming) could also subvert other forms of transmission. Research in this area—denying the enemy the use of their own radar or signals—was known as countermeasures, and the work was carried out in the Radio Research Laboratory at Harvard University with Dr. F. E. Terman (IRE President, 1941) in charge. The work was directed toward determining the vulnerability to jamming of various types of radio signals and designing very high-power transmitters for exploiting the weaknesses of enemy transmissions. Protection of one's own signals from enemy countermeasures, known as countercountermeasures, was also a task undertaken by the NDRC.

Among the countermeasures adopted were rapid shifts in the wavelengths used for radar and for field communications. Another was a passive shield known as "chaff." This consisted of great quantities of very thin metal strips of lengths suited to the wavelength of the enemy signals. These strips were thrown overboard by aircraft to form a screen and give false echoes for enemy radar. Today, chaff may be dispersed from warships to confuse oncoming radar-guided missiles.

Jamming is still used to interfere with short-wave broadcasts; it acts as a shield against adverse information and propaganda.

# TERMINATION
# OF THE WORK

*The innards of an early Loran transmitter, the R7.*

By the time the war ended, the Radiation Laboratory staff numbered 3000 scientists, engineers, and supporting staff, an extraordinarily able group. Disposition of the vast store of new science and technical know-how accumulated at public expense became a problem. The military argued against release of carefully guarded wartime secrets, but the potential aid to American industry outweighed their arguments.

It was decided, with Dr. Bush's help, that informed members of the Radiation Lab staff would stay on to prepare, at government expense, a full record of the Laboratory's technical work. This effort produced a monumental collection of technical literature — 28 volumes, published by McGraw-Hill from 1947 to 1953. The publisher agreed to make this heavy investment without assurance of return, but demand was brisk from the start. A sidelight is that the only monetary return to the U.S. Treasury on nearly $3 billion invested in the M.I.T. Radiation Laboratory came from several hundred thousand dollars of royalties earned by the *Radiation Laboratory* series of books.

# SPIN-OFFS
# IN PEACE TIME

In the work on jamming, it had been necessary to develop tubes capable of continuous outputs of hundreds of thousands of watts. Tubes of this nature, suited to operation on short wavelengths, were of great value in postwar development of television stations operating on channels 14 to 70. These ultra-high-frequency signals encounter greater absorption in the environment, and to make these channels competitive in coverage with the lower frequency channels 2 to 13, the FCC regulations permit effective radiated UHF signals in the range from 1 to 2 million W. The high-power work of the Radio Research Laboratory helped in this peacetime application.

Microwave radar principles are used in security systems, surveying, weather observation, and keeping track of earth satellites and space debris. The present density of air traffic could not be handled without the use of airport and enroute radar systems; planes are under surveillance at all points in their routes. Microwave radar is carried in all classes of commercial and private aircraft for weather observation, navigation, and collision avoidance. These routine radar procedures take on life-or-death dimensions in periods of bad visibility, when they serve as eyes for pilots and controllers. In marine service, radar is fitted on all classes of vessels, aiding in navigation and collision avoidance. Running at full speed through fog is now often practiced.

The microwave oven or "radar range" is powered by a cavity magnetron. The very short waves supply electromagnetic energy to the oven chamber. When the food is placed in the oven and the door closed, the radio energy penetrates the food and cooks it rapidly. The greatly reduced cooking time has made the microwave oven one of the most widely used home appliances.

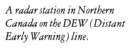

# SONAR:
# UNDERWATER "RADAR"

The word "sonar" is derived from "sound navigation and ranging." Sonar systems use underwater sound waves to detect the presence of ships, submarines, and other objects of interest. Submarines rely principally on passive sonar equipment that listens and tracks the noise created by engines and water disturbance by target vessels. Destroyers seeking submarines use active sonar, sending out pulses of acoustic energy and timing the echoes; submarines on attack also use active sonar. In both passive and active use, the devices that receive the echoes must be designed to indicate accurately the direction of the sound source.

The development of sonar began during World War I as a defense against the German submarine menace. Not much was then known about the acoustical properties of sea water, and during World War II there was established a U.S. Navy Underwater Sound Laboratory, operated by Columbia University in New London, CT. Dr. T. Keith Glennan was the director of this laboratory until the end of the war. Unlike the Radiation Laboratory, the Underwater Sound Laboratory has continued its work in the postwar period.

The speed of sound in sea water was found to vary with temperature, salt content, and many other factors. As a result, underwater sound waves travel in curved lines as water of differing salinity or density is encountered. These effects make sonar detection as much an art as a science.

The peacetime uses of sonar include navigation by reference to the contours and the depth of the sea bottom and the avoidance of submerged hazards. Seekers after sunken treasure use sonar equipment for discovery of sunken ships. A use of great interest to amateur and professional fishermen is in locating schools of fish. The sonar depth finder is standard equipment on all large ships and many smaller ones.

# ELECTRONIC AIDS
# TO NAVIGATION

In time of war, radar and sonar suffer the disadvantage of transmitting strong bursts of energy that disclose their source to the enemy; modern bombs and torpedoes use such signals for guidance to their targets. A system of navigation that required no radar or sonar emissions from the aircraft or ship was needed during World War II, and the answer came with loran.

The original idea belonged to Robert J. Dippy of the British Telecommunications Research Establishment, and it was named the GEE system. It was described by the British to the NDRC shortly after the Radiation Laboratory was set up; the responsibility for development was assigned to that laboratory. Long-range navigation was at first shortened to LRN, but this was soon changed to loran. The Loran Division was initially headed by the founder of the General Radio Company, Melville Eastham.

The GEE system used short waves to avoid sky reflections, but these could not be reliably received beyond the horizon. The task of the Loran Division was to make loran useful over much greater distances. John A. Pierce, who had joined the Radiation Laboratory from Harvard, suggested that waves of several hundred meters would be more useful for working over the horizon. There was the possibility that such waves would be reflected from the ionosphere, but the pulses carried over the earth's surface via the ground wave would arrive ahead of the reflected wave. If the pulses were very short, the ground wave could be distinguished readily from the sky wave and the latter ignored.

The successful part of Dippy's scheme involves precisely timed radio pulses transmitted from two accurately located stations — a "master" and a "slave." By an adjustment on his receiver, the navigator superimposes the received pulses from the two stations on a cathode-ray screen. By reading the dials, the difference in the time at which the two pulses arrive at the receiver is measured; using the known speed of radio waves, the time difference is transformed to a difference in distance.

The navigator then knows he is located at the apex of a triangle, the base of which is the line connecting the two stations, and from his loran receiver he also knows the difference in length of the other two sides. The locus of the apex of all triangles having the measured difference is a hyperbola, plotted on his loran map. He then repeats the measurement using pulses from his original master and a differently located slave station, thus placing himself on a second hyperbola on his map. His ship or aircraft is located at the intersection of the two hyperbolas.

The first loran transmitter was installed at a Coast Guard station in Ocean City, MD, in November 1941. Its 100-kW pulses began to ring the bells of

*Top: A Loran transmitting station on the Faroes Islands in the North Atlantic.*
*Below (left and right): The first mobile two-way FM radios were built and installed by Motorola in 1940.*

the ship-to-shore telephone service on Great Lakes ore carriers, but after Pearl Harbor the loran service was moved to the then-vacant 160-m amateur band. This was the loran A service, not closed down until 1981. The postwar service, loran C, uses wavelengths of 3000 m that can span greater distances. The power of the loran C pulses is about 5 million W, radiated from towers 1300 ft high. To secure positions within a few hundred feet at distances up to 1600 km (1000 mi), the pulses are matched by the individual oscillations of the respective radio waves.

# MOBILE
# COMMUNICATIONS

The military importance of avoiding wire lines by use of radio communications was well appreciated by all the powers after World War I. Between the wars much progress was made, most based on civilian developments. Teletypewriters replaced the telegraph key, and the radio amateurs' exploration of the short-wave spectrum opened up new channels for long-range military communications. A new approach to mobile communications appeared in the civilian field in the late 1930's, when Prof. Daniel E. Noble of the University of Connecticut decided that Armstrong's noise-defeating frequency modulation system would be well suited to communi-

*The first Walkie-Talkie units were used in the Allied invasion of Italy. They rapidly became an indispensable link in military communications.*

cations between police cars and headquarters. The Connecticut State Police tested his ideas; the tests were a success, and soon the method was extended to fire departments and taxicabs. A major new business appeared for Motorola, and Dr. Noble joined the Motorola staff; under his direction mobile radio made rapid progress.

But mobile equipment was not portable equipment. It was still too heavy to be carried by the foot soldier because it required a heavy storage battery for heating the vacuum tube filaments. Development of a small tube with reduced power requirements for its filament allowed some progress. During the war, Dr. Noble led in the design of a truly portable combined receiver and transmitter (transceiver), leading to the Army's Walkie-Talkie. This equipment gave a range of 1.6 km (1 mi) and was widely used in battlefield situations.

The wartime Walkie-Talkie became the peacetime Handie-Talkie, now familiar to trainmen, construction workers, policemen, and firemen, when quick access to information is required. Today's Handie-Talkie uses transistors and operates at shorter wavelengths than its wartime relative.

## FROM MISSILES
## TO SATELLITES

When the Russians launched Sputnik in October 1957, the age of the earth-girdling satellite began. The U.S. countered with a satellite early the next year. It was a direct result of the work done under Dr. Werner von Braun that had produced the V-1 and V-2 rockets for Germany at Peenemunde on the Baltic Sea. These were Hitler's secret weapons, but they were defeated by the British because the gliding bombs kept a set course, making them ideal targets for radar-tracked antiaircraft guns. At the end of the war, von Braun was brought to the Army installation at Huntsville, AL, where he developed early U.S. space missiles.

The NASA space program reached a high point on July 20, 1969, when

Astronauts Neil Armstrong and Edwin Aldrin landed on the moon and then safely returned to earth. The worldwide television coverage of this event gave clear evidence of the central role being played by electronic systems. Electronic methods used in space flight monitor each astronaut's physical condition and the conditions on board the space vehicle, with signals carried by radio telemetry to the command center. Computers on board and on earth accurately control each flight.

More recently have occurred the fly-bys of Mars and the unmanned landings there, and the remarkable series of photographs of Saturn and its rings and satellites. Still to come may be photography from the outlying planets of the solar system, as the Saturn spaceship approaches these planets.

# SATELLITES
# THAT TALK BACK

The first proposal for relaying information by radio to and from an artificial earth satellite was made by science fiction writer Arthur C. Clarke in the October 1945, issue of the British journal *Wireless World*. He pointed out that a satellite at an altitude of 36 000-km (22 300-mi) would have an orbit synchronous with the Earth's rotation and hence would appear to maintain a fixed position in the sky. Clarke proposed that radio signals be beamed to such a "geostationary" satellite, and be retransmitted to earth. Signals from three such satellites could cover the earth.

Reflection of signals from passive objects in the sky had been achieved earlier. In fact, in early 1946, John D. DeWitt, Jr. and his associates located at Evans Signal Laboratory, Belmar, NJ, succeeded in detecting 2.7-m pulses reflected from the moon's surface, using a large radar antenna. In 1960, the U.S. Navy utilized moon-reflected communications from Washington, DC to Pearl Harbor, HI. In August 1960, ECHO, a large metal-coated plastic balloon, was orbited in a package and inflated by internal gas pressure. It orbited 1000 mi

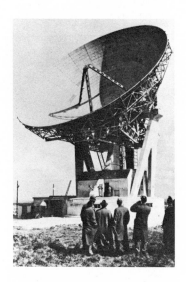

*TELSTAR, although launched by the United States, had ground tracking stations around the globe. This one was at Goonhilly Downs, Cornwall, England.*

above the earth and was used as a signal reflector.

The first active (relaying) satellite was TELSTAR, launched in July 1962, by NASA. Designed and built by AT&T, it was used for overseas communications and experiments. TELSTAR was nonsynchronous, having been too heavy to send into synchronous orbit with the launchers of 1962. Using a ground station at Andover, ME, it carried live transoceanic television programs for the first time from the U.S. to Europe, South America, and Japan. Equipment to match the varying worldwide television standards was an achievement in itself. The next year, on July 26, 1963, NASA launched the first synchronous satellite, SYNCOM.

In April 1965, the International Telecommunications Satellite Consortium, a cooperating group of 11 member nations which has now grown to 109, launched EARLY BIRD (INTELSAT I) from Cape Kennedy as the first sponsored fixed-position satellite. Six years later, technical progress achieved the launch of INTELSAT 4, to handle 6000 duplex telephone circuits. It weighed 696 kg (1534 pounds) compared with the 39 kg (85 pounds) of INTELSAT I. By 1981, satellite relay of intercontinental programs was so routine as to go unremarked by the announcers describing the wedding of Prince Charles and Lady Diana, televised to a worldwide audience of several hundred million viewers.

## EYES IN THE SKIES

Another use of satellites of both military and civilian importance is observation of the earth and its atmosphere. Pictures of the earth and its cloud cover are used routinely in evening weather forecasts. Other uses include mapping of terrain, geological studies, monitoring of agricultural areas and crop condition, and archeological discovery.

For these uses the orbiting satellite carries a still camera capable of very fine detail, with images electrically recorded on a magnetic disk. Each orbit ranges over the earth several times a day, and when the satellite passes over its ground station it transmits or "dumps" picture signals derived from its magnetic records. Combined with a computer, the received pictures can be enhanced or shaded to separate various areas of interest with contrasting colors.

The detail of these pictures is astounding to the uninitiated. Such pictures have military as well as civilian use, keeping track of the erection of military installations and the deployment of weapons. Cameras sensitive to infrared light can detect crop diseases. They can also detect the movement of earth in military construction by the heat generated in moving it. The satellite and its cameras are powerful tools in monitoring adherence to arms reduction and control agreements, important in the quest for a lasting peace.

It has been foreseen that before 1990 technical progress will permit direct reception of satellite television programs at homes equipped with antennas less than 1 m (3 ft) in diameter. Also required will be frequency translation equipment to change the satellite wavelength of about 25 cm to the longer waves of the domestic television band.

Achievement of this objective raises an international problem, that of confining satellite transmissions to the boundaries of individual nations, particularly in Europe. This objective was agreed to at an international conference in 1973, and raised severe problems in satellite antenna design. Moreover, there will always be some stray signal, and with more sensitive receivers the opportunity to eavesdrop across borders will certainly occur. Whatever corrective action is then taken falls in the political arena; Maxwell's laws cannot be repealed.

Two other problems are evident on the horizon: crowding of satellites in geosynchronous orbit and crowding of satellite frequencies in the available spectrum space. A "parking orbit" at 36 000 km (22 300 mi) having a circumference of 266 000 km (165 000 mi) seems to provide sufficient separation for numerous satellites, but those parts of the parking circle over centers of high population are more crowded than others. If satellites are parked too close to each other, interference between upcoming signals may occur, restricting the choice of usable frequencies.

More pressing is the choice of operating frequencies in the bands assigned to the satellite service. Direct satellite broadcasting to the home will require a band of 27 MHz, more than four times as wide as the channel used for earthbound stations in North America. If every country insists on its own satellite service, with three or more channels each, the satellite bands will be heavily oversubscribed. Shorter waves might be used but such signals are weakened when they pass through areas of heavy rainfall. Again, the contest for the future use of communications satellites will involve political as well as technical progress.

Can there any longer be doubt that the new horizons of their field require that electrical engineers roam far afield?

## For Further Reading

D. G. Fink, *Radar Engineering*.   New York: McGraw-Hill, 1947.

E. L. Ginzton, "The $100 idea," *IEEE Spectrum*, p. 30, Feb. 1975.

D. Hanson, *The New Alchemists*.   Boston, MA: Little, Brown, 1982.

R. K. Jurgen, "Captain Eddy: The man who 'launched a thousand EE's,'" *IEEE Spectrum*, p. 53, Dec. 1975.

E. C. Pollard, *Radiation—Essays on People and the Work of the MIT Radiation Laboratory*.   Durham, NC: Woodburn, 1982.

Special Issue on Technology in War and Peace, *IEEE Spectrum*, vol. 19, Oct. 1982.

# 9

# ALL THE WORLD BECOMES A STAGE

$A$s with much of electrical engineering, the story of television has roots extending back to the decades before AIEE was established. Although late nineteenth-century science provided basic knowledge, early attempts at picture transmission failed for lack of suitable materials and apparatus. In fact, no one was able to transmit any form of picture until about 1925, after electronic amplification, photoelectric cells, and the neon lamp were added to the Nipkow disk of 1884.

It was understood that many requirements had to be met to transmit pictures of motion to distant locations. In some fashion, the scene had to be broken into small "picture elements" of differing brightness; the brightnesses of these elements had to be translated into corresponding electric currents; these currents had to be transmitted to the receiver, converted back into light, and presented to the eye; the picture elements viewed at the receiver had to be in the same arrangement and of the same relative brightnesses as those at the transmitter; and a succession of pictures had to be sent at a rate high enough to avoid flickering.

## BASIC SCHEMES

$I$n 1879, George R. Carey of Boston and Denis D. Redmond of Dublin proposed a multiwire scheme, wired element-to-element. But the number of circuits had to equal the number of picture elements. Today that would call for about 200 000 separate circuits. Obviously, such parallel transmission of picture information would be too complex and expensive.

*Polly Thompson and Helen Keller, who is holding her hand over the speaker of a television set made by Donald Fink.*

An alternative would be to send the currents associated with the picture elements one at a time over a single circuit. This would require exact synchronizing of the successive views of the elements at transmitter and receiver. This scheme has been the basis of all television systems from 1925 to the present. In an early exemplification of this idea, Alexander Bain, a Scottish watchmaker, in 1843 suggested a telegraph which would transmit the characters for messages by producing stains on chemically treated paper. The concept embodied all the geometrical and timing methods of the modern television system.

# EARLY IDEAS

Two major discoveries were needed to implement a system that would fulfill these requirements: a device to translate light into electrical current, and another device to translate the current into light at the receiver. The latter invention was the last piece to fit into the puzzle. It produced a practical television system.

In 1873, Willoughby Smith and his assistant, Joseph May, working on the Atlantic telegraph cable, noted that the element selenium changed its electrical resistance when light fell on it. This may have been only a happy accident, but it gave a first solution for the translation of light into current.

By 1878, Prof. Adriano de Paiva of Portugal suggested an electric telescope in which an image was to be focused on a selenium plate that was scanned by a moving stylus. At the receiver a lamp was to move in synchronism, its brightness changing with the varying current from the selenium.

Also in 1878, a French lawyer, Constantin Senlecq proposed a transmitter of small selenium cells. By means of a commutator, each cell was to be connected serially to a single wire to the receiver. Through a similar commutator, the signal was to be passed to an array of cells of fine platinum wire, each wire to glow with a brightness proportional to the current in the selenium cell.

Other proposals were made for using selenium, but all were beyond the state of the existing mechanical and electrical art. They also did not recognize that the change in resistance of selenium, and many other photoconductors, lags behind the change in light intensity. Only in 1956, when the RCA vidicon camera tube was introduced with antimony sulfide as the photoconductor, was this sluggishness overcome.

A better method of translating light into electricity was the photoelectric effect, discovered in 1889 in Germany by Julius Elster and Hans Geitel. They found that when the surfaces of certain metals — such as sodium, potassium, rubidium, or cesium — were illuminated in a vacuum, they assumed a positive charge. Today, we know that the surfaces emit electrons.

This effect was explained in 1905 by Albert Einstein, when he proposed that the light falling on the photoelectric plate must fall as individual packets of energy, later named photons. If a photon has sufficient energy, it can liberate an electron from the metal surface. Emission occurs almost instantaneously upon illumination of the surface. Einstein's photoelectric law and its idea of quantized light supported the earlier work of Max Planck on thermal radiation of energy. In 1913, Niels Bohr extended the concept in his theory of electron behavior in the atom, which was the keystone for our understanding of electrical conduction, particularly in the transistor.

But there was still the problem of translating current into light at the receiver. In 1917, D. McFarlan Moore, a G.E. engineer, produced a lamp containing neon gas and giving the familiar red glow when an electric current passed through it. Variation of current produced a change in intensity of the glow, as was needed for showing television images.

## THE NIPKOW DISK

The Nipkow disk was a device for scanning a scene synchronously at the transmitter and receiver, point-by-point.

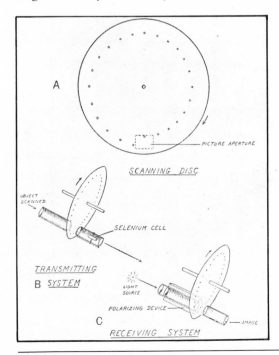

A schematic diagram of the Nipkow disk system.

Patented in Germany in 1884 by Paul Nipkow, it consisted of a flat circular plate through which small holes were drilled along a spiral line near the periphery. The disk was placed between the scene to be scanned and a light-sensitive cell. As the disk was rotated, the first hole traced a line across the scene and transmitted light to the cell. The second hole followed but, being on a spiral line, it traced out a line of the scene below the first; the process was repeated by each successively lower hole. The total scene was thus scanned, in the time of one revolution, into as many lines as the disk had holes.

The Nipkow disk lay dormant for 40 years, awaiting development of missing links in the art, namely, the amplifier and a rapidly variable light source, both finally available by 1920. By that time it was possible to increase the output of the photoelectric cell by an electronic amplifier, and to modulate the light output from a neon glow lamp. The receiver included a second disk. Behind it was a neon lamp having a rectangular plate covered by a red glow. The transmitted signal caused the light to vary, and the changing light emerging from the holes in the disk could, with some imagination, be seen as tracing the corresponding lines of the transmitted scene. If the rate of disk rotation was at least 15 revolutions per second, the image did not appear to flicker. Synchronous motors were available to keep the two disks in step, provided that the power at transmitter and receiver was supplied by the same system.

Two improvements to the Nipkow system then followed. First, since the small holes in the disk could pass only a tiny part of the light reflected from the scene, the holes were enlarged and fitted with lenses to pick up more light and focus it on the photocell. In another approach, a prismatic glass ring replaced

*The scanning apparatus used by Charles F. Jenkins.*

the holes, the angle of the prism changing continuously around the edge of the disk. When light passed through this ring as the disk rotated, the light paths were swept downward. By use of a second ring, the light paths were also swept horizontally. In another variation of the Nipkow scheme, the "flying spot" technique, an intense beam of light behind the transmitting disk passed through the holes and illuminated the scene with a moving spot of light. The varying light reflected from the scene was picked up by a bank of photocells.

## FIRST DEMONSTRATIONS

Two workers employed the Nipkow scanning principle in early television systems. On June 13, 1925, Charles F. Jenkins, born in Ohio in 1867 and a worker with motion picture cameras, gave a public demonstration of his apparatus. The signals were transmitted from the Navy station NOF at Anacostia, near Washington, DC. He used the flying spot method of pickup with a lamp behind prismatic rings and a glow lamp and prism disk at the receiver. His received picture appeared on a reflecting screen measuring about 15 × 20 cm (6 × 8 in). The images were coarse, being scanned with only 48 lines.

Preceding Jenkins' demonstration by three months, in March of 1925, the 37-year-old Scotsman John Logie Baird showed his television system in Selfridge's department store in London. He had begun his work on television in 1923 in a garret in Hastings, England. His Nipkow disk was fitted with specially arranged lenses. The images were scanned in only 8 lines, but each horizontal line was composed of 50 elements, giving 400 picture elements. Initially, both Jenkins' and Baird's demonstrations produced only silhouettes, although they shortly improved their systems to reproduce halftone images.

Both men were aware of the value of publicity, and there was a flurry of promotions and broadcasts. The British Broadcasting Corporation (BBC) began experimental broadcasts using Baird's equipment, with 30-line images, in September 1929.

Ernst F.W. Alexanderson at G.E. entered the field in 1928 with a 24-in Nipkow disk that scanned at 48 lines per image. In September, he used a portable Nipkow camera to televise New York Governor Alfred Smith's acceptance speech for the Democratic nomination for President, delivered on the State Capitol's steps in Albany.

## AND IN COLOR, TOO!

In 1928, in England, Baird transmitted color images using a Nipkow disk with three sets of holes that scanned the scene through three color filters. The three lines were scanned sequentially, a technique later used in all-electronic television by Peter Goldmark.

In June 1929, Herbert E. Ives and co-workers of Bell Laboratories also demonstrated color transmission, using the flying-spot system with three photocells fitted with orange–red, yellow–green, and green–blue filters. Three separate wire circuits transmitted the signals to the receiver where they controlled three gaseous lamps: a neon lamp with a red filter, an argon lamp with

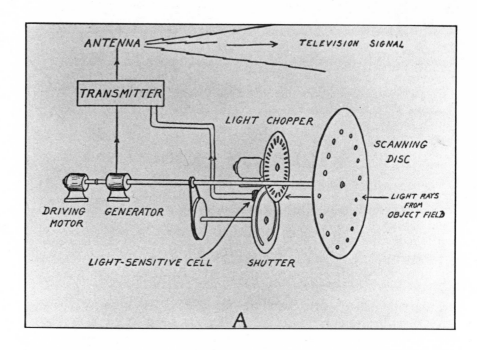

a blue filter, and another argon lamp with a green filter. The light from these lamps passed through the apertures of a Nipkow disk and the three colored images were superimposed on a tiny screen.

The colors of an actor's face, a watermelon, the U.S. and British flags, and a woman's dress were clearly evident in coarse detail with 50 lines per picture. The demonstration was a promising augury of things to come.

# THE END OF
# MECHANICAL SYSTEMS

By the end of 1928, there were 21 stations licensed in the U.S., with 15 stations on the air. Various frequencies were used, some in the standard broadcast band and some in the short-wave bands. There were no standards for the number of lines or number of pictures per second used by these stations, and no attention was paid to the bandwidth required to carry the transmissions.

The lack of standardization was clearly an impediment to the sale of receivers to the public. In 1928, the Radio Manufacturer's Association (RMA) established a television standards committee, and in October that committee recommended scanning the lines from top to bottom of the scene, and along the lines from left to right. Also recommended was a picture repetition rate of 15 frames per second, with 48 or 60 lines per frame. Today's television broadcasts use scanning rates of 25 or 30 frames per second, originally related to the local power frequencies, and the number of lines is either 525 or 625 per frame.

It was soon recognized that the mechanically scanned systems could not produce pictures of sufficient detail to meet the expectations of the public. As a result, the television boomlet ended in 1937. The crowning defeat occurred

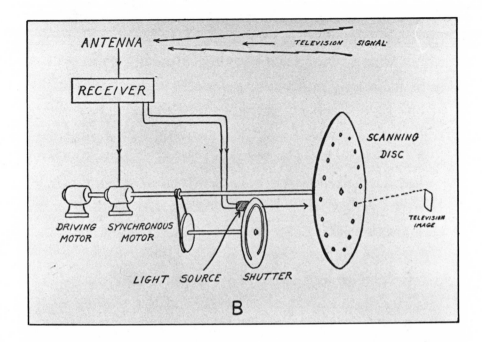

in the public transmissions inaugurated in London in 1936. The BBC began a broadcast service using an improved Baird system with 240 mechanically scanned lines. For distant scenes, this system involved a 40-s processing of motion-picture film, followed by mechanical scanning of the film frames. The BBC also broadcast programs using the Marconi-EMI electronic system, which had no moving parts other than the electrons, produced 25 frames per second and scanned 405 lines per frame. The Marconi-EMI camera was much more sensitive in dim light than the Baird mechanical scanner, and it was not limited to a small number of lines.

From the beginning of the British tests it was clear that the electronic system was superior to the mechanically scanned one, and the BBC decided in February 1937 to standardize on the Marconi-EMI 405-line system. Meanwhile, a fire had destroyed much of the Baird equipment being used to develop an electronic system based on the inventions of the American Philo T. Farnsworth. The abandonment by the BBC of the Baird transmissions ended the mechanical era of television service.

# THE DAWN OF
# ELECTRONIC TELEVISION

The principles of a system of electronic television were conceived early and progressed with the art. Available during the period prior to 1937, but still too crude to be effective, was the Braun cathode-ray tube. Invented in 1879 by Ferdinand Braun of the University of Strasbourg, it employed a "gun," producing a beam of electrons, located at the narrow end of a glass funnel. The inside surface of the glass enclosing the wide end of the cone was coated with a fluorescent material that showed a spot of light where the electron beam struck. The beam could be deflected over the fluorescent screen by varying voltages applied to pairs of perpendicular internal plates, or by magnetic fields. This device was to become the kingpin of the all-electronic television system.

Two German patents of 1906 and 1907, to Max Dieckmann and to Boris Rosing, the latter of the Technological Institute in St. Petersburg, proposed the use of the cathode-ray tube for reception, with the Nipkow disk scanning the televised scene. Rosing proposed modulating the electron beam, and thereby the brightness of the spot on the screen, in accordance with signals coming from the scanner. The beam of the Braun tube was to be deflected synchronously by two electromagnet coils on the neck of the tube, in line with today's practice. Rosing produced a crude image in 1911, but his work was interrupted by World War I, and he disappeared in the ensuing Russian civil war.

Rosing is now the hero of Russian television history, but his major impact on the development of electronic television elsewhere came through one of his students at the Institute in St. Petersburg — Vladimir K. Zworykin. Zworykin was born in 1889 in Mourom, Russia, and came to the U.S. in 1919 to escape the upheavals in Russia. He first went to work for Westinghouse in Pittsburgh; he left that company in 1922, but returned in 1923. In that year he applied for a patent on an all-electronic television system.

However, the first proposal for an electronic system had been put forward by A. A. Campbell-Swinton, an English physicist, in a letter to *Nature* in 1908. Although impractical then, several of his suggestions anticipated important television components of a later day. His great contribution was to propose use of a form of Braun tube at the transmitter. This proposed tube contained a flat insulating plate, on one side of which small, photoelectrically sensitive rubidium cubes were deposited, insulated from each other. The image to be televised was focused on this array of tiny photoelectric cells, causing them to lose electrons in proportion to the amount of light falling on each cube. The electron beam, scanning the reverse side of the plate, would restore the electric potential of each cube, and the resultant current would be a measure of the change

*The young Vladimir K. Zworykin.*

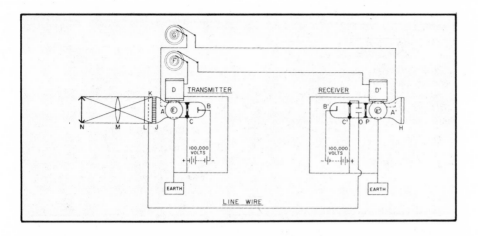

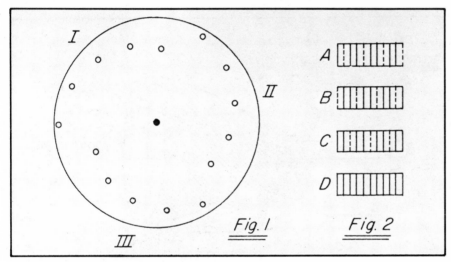

*Fig. 9.1 (top): A. A. Campbell-Swinton's proposal for an all-electronic television system. Fig. 9.2 (left): Ulises A. Sanabria's triple-spiral Nipkow disk.*

in brightness from cube to cube. This proposal was an accurate prediction of the iconoscope, the electronic camera tube later proposed and developed by Zworykin.

Most important was a basic principle: during the full time between scans, each cube would continue to lose electrons at a rate in proportion to its illumination, thus storing the electric charge. This effect greatly increased the sensitivity of the camera tube to light. Diagrammed in Fig. 9.1 is Campbell-Swinton's system, one of the landmarks of electronics.

Campbell-Swinton's system failed in its day because the available Braun tubes were crude and no one knew how to produce the mosaic of rubidium cubes.

Alexander M. Nicholson of Western Electric Laboratories supplied another essential principle in a patent application of 1917. His mechanical scanner, an oscillating mirror, produced not only the picture signal for each line, but also another signal for synchronizing transmitter and receiver. This was the first composite signal, embodying both picture and timing information.

In 1929, Ulises A. Sanabria of Chicago proposed a new type of Nipkow disk

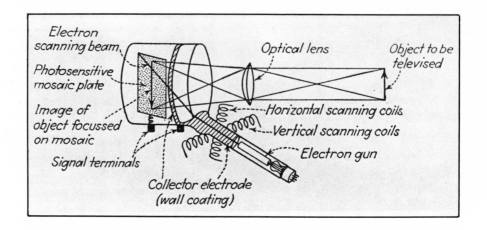

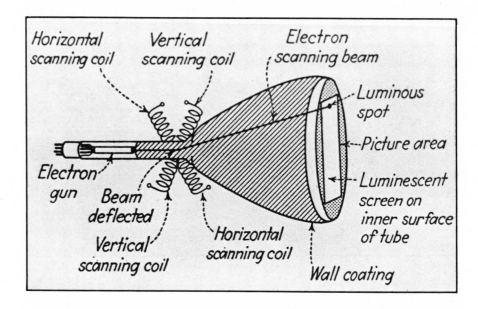

(Fig. 9.2) with the spiral of holes arranged in segments. The two-segment version, known as interlaced scanning, is now standard throughout the world.

# ALL-ELECTRONIC
# TELEVISION ARRIVES

At the heart of the Zworykin patent proposal of 1923 was the iconoscope, a charge-storage tube similar to that suggested by Campbell-Swinton. The image plate was covered with aluminum oxide, on which was sifted a potassium compound. By heating, the powder was melted into tiny globules of potassium hydroxide, each insulated from its neighbors. These were the equivalent of Campbell-Swinton's photosensitive rubidium cubes.

In Zworykin's original proposal, the plate was installed in a vacuum tube and its reverse side was scanned by the electron beam. The charge on each globule continually grew until it was neutralized by the electrons of the beam. The output current was proportional to the light reaching each point. This signal was amplified and applied to control the beam intensity in another form of Braun tube, called a kinescope. Synchronizing signals were used in the beam deflection, as Nicholson had proposed.

Zworykin's demonstration to Westinghouse officials in 1923 was disappointing, but work continued. In 1930, he was transferred to the reorganized RCA, and by 1933, he had improved versions of iconoscope and kinescope ready for demonstration. The iconoscope now employed direct impingement of the electron beam on the globules, the discharge current being picked up from a metal coating on the reverse side of the image plate.

In later years, there followed improved versions of the iconoscope — the image iconoscope, the orthicon, and the image orthicon — each more sensitive than its predecessor and more free of internal noise.

*Zworykin with an iconoscope.*

# SARNOFF
# AND FARNSWORTH

In 1933, RCA began field tests of the Zworykin system from a transmitter at its plant in Camden, NJ, using 240-line scanning. The next year, it introduced the interlace principle of Sanabria and went to 343 lines with 30 pictures per second.

These tests convinced David Sarnoff, head of RCA, that the time for commercial exploitation of electronic television was at hand. In May 1935, he announced RCA's plans to invest $1 million to place a transmitter atop the Empire State Building, to install studios at Radio City, and to develop receivers suitable for use by the public. Also developing television plans were the Dumont Laboratories in New Jersey, the Farnsworth Laboratories in San Francisco, G.E. at Schenectady, Philco in Philadelphia, and Zenith in Chicago. Well aware of the lead held by RCA, they wanted their places in the television sun.

RCA found in 1936 that a major blockade was ahead in the form of a patent held by Philo T. Farnsworth. Farnsworth was born in Utah in 1906. Without a college degree, although he attended classes at Brigham Young University, he was the archetype of the lone inventor. For television he had invented a device described as an "image dissector." The image to be transmitted was focused on a photoelectric coating on a plate in a cylindrical vacuum tube. The coating emitted electrons point-by-point in proportion to the light, thus forming an image in electrons that was a precise copy of the optical image. A positive electrode at the opposite end of the tube drew these electrons away from

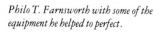
*Philo T. Farnsworth with some of the equipment he helped to perfect.*

*The picture on this set was viewed in a mirror on the underside of the raised lid.*

the coating, and a magnetic field kept the electrons in their proper positions (in focus) as they passed down the tube.

Additional magnetic fields moved the electron image bodily, horizontally at a fast rate, vertically at a slower rate, effectively scanning the electron image past a small aperture. Electrons entering this aperture constituted a current that was amplified and sent to the receiver.

The dissector tube did not employ the charge-storage principle, but by 1938, Farnsworth had developed image dissectors in which that principle was used.

Zworykin's patent on electronic television, applied for in 1923, was not issued until 1938 because numerous patent interferences were filed. One of these came from Farnsworth, as a result of which the patent examiner ruled that Zworykin could not make the claim of scanning an electron image. RCA's appeal was denied, and it appeared that RCA would have to come to an agreement with Farnsworth in order to proceed further with its iconoscope development. Farnsworth would not sell his patent rights and insisted on a licensing agreement, which was contrary to past RCA policy. Finally, in 1939, RCA and Farnsworth did sign a nonexclusive cross-license agreement, which also covered several other inventions having value for both parties. RCA was then in position to proceed with its grand plan to commercialize television.

## STANDARDS AGAIN

Once more RCA's plans were obstructed, this time by the Federal Communications Commission (FCC). In 1935, an RCA demonstration of its 343-line all-electronic system to members of the Radio Manufacturer's Association had been impressive. Believing in the imminent possibility of a public service, RMA instructed its Engineering Department to determine the feasibility of establishing standards for television broadcasting.

The FCC, hearing of this action, announced that it would hold hearings, starting in July 1936, to set up policies governing the allocation of radio frequencies for television stations. The FCC Chief Engineer, T. A. M. Craven, wrote to RMA suggesting an industry agreement on television standards prior to action by the FCC. He stated the policy that any television receiver manufactured for public use must be capable of receiving transmissions from any station licensed by the Commission.

RMA established two technical committees, one on channel allocations and one on technical standards, and instructed both to furnish a report for the 1936 FCC hearings. The work of these committees underlies much of today's television broadcast practice in the U.S. Recommended was a 6-MHz channel, which remains the standard. Today this seems restrictive by comparison with the 8-MHz channels adopted much later in Europe, but in 1936 it was visionary.

The other RMA committee recommended an image of 441 interlaced lines, scanned at 30 pictures per second, and recommended frequency modulation for the sound channel. These standards remain, except for a 1941 increase in the number of lines to 525. Following the 1936 hearing, the FCC allocated 19 6-MHz channels in the spectrum from 44 to 294 MHz.

In December 1939, the FCC stated that it was prepared to authorize transmissions with "limited commercialization" (sponsored programs), with the income to be used for experimental development of program service. It did not act on the RMA proposals, and in fact warned against any broadcaster attempting to "freeze" on any standards without further Commission action.

# THE FIRST NTSC

The opening of the New York World's Fair in April 1939 proved an irresistible challenge to David Sarnoff of RCA. He used the occasion to demonstrate a complete system, using 441-line standards,

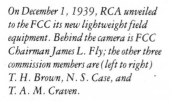

*On December 1, 1939, RCA unveiled to the FCC its new lightweight field equipment. Behind the camera is FCC Chairman James L. Fly; the other three commission members are (left to right) T. H. Brown, N. S. Case, and T. A. M. Craven.*

*What the television of the 1930's lacked in picture size, it made up in ostentation and material. Notice the matched woods in this cabinet.*

not yet Commission-approved. He also announced plans to produce receivers in quantity for the public; in fact, production had started in January.

The reaction of the FCC was swift. Urged by objections from RCA's competitors, who offered different standards, the FCC, in May 1940, withdrew the limited commercial authorizations, stating that RCA's action had established a de facto set of standards without Commission approval.

With commercial operation stopped, not to be resumed until industry agreement on standards was reached, the time was ripe for the formation of the first National Television Systems Committee (NTSC). This committee recommended standards for monochrome television that were adopted by the Commission, and commercial operation was authorized, effective July 1, 1941. The only major change in the previous standard was the increase in lines from 441 to 525.

The first NTSC was the joint creation of FCC Chairman James L. Fly and Walter R. G. Baker, a G.E. executive who was also serving as head of the engineering department of RMA (IRE President, 1947). Fly asked Baker to form an all-industry committee that would break the standards impasse, and Baker responded in masterly fashion. Inviting all competent engineers to deal with the issues, whether or not employed by RMA member companies, he devised a panel structure and parliamentary procedures that would draw attention to self-interest and force individuals to concentrate on the joint interest of the profession and the industry. In nine months, the work was done by 168 committee and panel members. All standards proposals were subjected to review and argument, with defense of different backgrounds and positions. Thanks to NTSC, monochrome television got off to a flying start.

The December entry of the U.S. into World War II put an end to further development of domestic service. Stations were allowed only limited hours of operation, but the scene was set for a rapid postwar revival.

# INCOMPATIBILITY
# IN COLOR

John Baird and Herbert Ives had demonstrated color television using mechanical scanning in 1928 and 1929. In the electronic era, work on color television was undertaken in the 1940's by Peter C. Goldmark of the Columbia Broadcasting System. He arranged to scan the picture three times through red, green, and blue filters. These were arranged as segments of a disk which rotated in front of the camera, in synchronism with the picture scanning. In this way, three fields were successively scanned: the first was formed with red light, the second with green, and the third with blue. This was known as a field-sequential color signal.

At the receiver, the signal was reproduced on a black-and-white picture tube that was viewed through a similar rotating disk having three filter segments. The disk was synchronized so that the red filter intervened when the red filter was on the camera, and so on. Thus, three pictures were required for color, whereas one was supplied for a black-and-white picture.

If the monochrome rate of 30 pictures per second was maintained, then 90 pictures would have to be scanned each second for color. Since the width of the frequency channel is proportional to the number of pictures scanned per second, color would require three times as much frequency space as monochrome. Since no such range was available in the frequency spectrum, Dr. Goldmark compromised, reducing the color-sequence scanning rate to 60 per second and the scanning lines from 525 to 343 per frame.

But the scanning rate for 343-line, 60-frame images differed radically from the 525-line, 30-frame monochrome standard in public use. The scanning rates of color and monochrome systems were inevitably incompatible. Receivers could have been designed to accommodate both scanning standards at some increase in cost, but the hard fact was that the receivers already in the hands of the public had not been so designed.

# COMPATIBILITY
# IN COLOR

What was needed, and what was eventually achieved, was a system of color television that used the same scanning rates and channel as the monochrome service; that is, a color system compatible with the black-and-white system.

Goldmark first demonstrated his 343-line, 60-frame per second sequential color system to the press in September 1940. Those of the NTSC present were impressed, and in their report to the FCC, they recommended authorization of experimental color telecasts. But at that time, no one foresaw the possibility of a compatible color system. How could one transmit three times as much information in the channel designed for monochrome service?

The rapid postwar growth of television service almost exhausted the broadcasting channels reserved by the FCC. So, in September 1948, the FCC stopped issuing licenses for new television stations; it did not relax the freeze until April 1952. During those years, the feasibility of a compatible color service was established.

An answer to the problem was found by concentrating on the basic fault of the field-sequential system: it gave the eye more information than it could use. Specialists in human vision had long known that the eye can discern fine detail in white light, somewhat less detail in green, still less in a red image, and much less in blue. Goldmark's system offered equal detail in all three colors. It became clear that if the detail offered were reduced to match the eye's abilities, less information would need to be transmitted and a saving in channel bandwidth could be made.

RCA had been working on a color system using three channels and described it in 1947, but the lack of frequency space made the effort futile. The freeze order of 1948 was the signal for David Sarnoff to issue a challenge to his team of talented television engineers: invent a color system that would fit into one 6-MHz channel and that would provide a monochrome picture on the monochrome receivers then in use. At that time there were nearly two million receivers in use in the U.S., with new receivers reaching the public at the rate of 400 000 per month. Sarnoff saw the need to preserve the value of those receivers.

In 1949, CBS petitioned the FCC for authority to broadcast color programs scanning at 405 lines, 72 pictures per second, with the signal confined within the limits of the 6-MHz channel. Several groups knew that a compatible system was in development, and on August 25, 1949, RCA broke its silence and described its compatible color system, with a complete set of standards.

The contest between the CBS incompatible system and compatibility came to a head in an FCC hearing that began in September 1949. It was held to establish the practicability of transmitting a color signal on a 6-MHz channel. Prior to the hearing, the Joint Technical Advisory Committee (JTAC) assembled the available information for the Commission. During the hearing, the Commission had to digest such a mass of conflicting evidence that it could not see the forest for the trees. It turned down the compatible proposal and accepted CBS's incompatible system. RCA fought this result to the Supreme Court, which ruled in favor of the Commission.

So, on June 21, 1951, CBS color transmissions began in New York. There were few color receivers available and none of the hundreds of thousands of monochrome receivers could be adjusted to receive the color transmissions. The CBS effort ceased some four months later, due to the limitation on receiver production that came with the Korean War.

The era of good feeling between the FCC and the television industry had perished.

# NTSC II

When the potential for compatible color television became evident in 1949, Dr. Baker recalled the NTSC, and by January 1950, it was ready to hold its first meeting. Needless to say, the FCC was not pleased with the action. In October of that year, the Commission declared that compatible color was not good enough to be adopted for public use, and FCC engineers were not permitted to attend NTSC meetings until 1952. By that time, the CBS system had been abandoned.

The second NTSC had a vastly more complex job than did the first committee. The technical issues were close to the limits of technology. Color television's basis is that any scene may be represented by superposition of three identical images, as seen through red, blue, and green filters. When such images are presented to three camera tubes, the resultant signals may be combined in several ways. If they are added electronically, the result represents the scene in black and white, and this combined signal is, in fact, transmitted in the NTSC color system. On the color screen, this combined monochrome signal contains all the fine picture detail. It also provides a picture on monochrome receivers.

Additional information on the color content of the scene has to be transmitted for color receivers, but it must be ignored by monochrome receivers. A key factor in the NTSC signal is that the additional color information need not contain fine detail. The situation is reminiscent of the days before color photography, when water colors were painted over a black-and-white print, and only broad strokes of the brush were needed to convey realism. In color television, an equivalent effect is achieved by transmitting colors in coarse structure. Since less information is sent, less bandwidth is required.

We have two images to deal with: the monochrome image produced by the combined color signals and the full-color image we desire. The color information is manifestly the difference between these two images. In other words, it is only necessary to transmit color-difference information in addition to the monochrome image. Electronic subtraction is as easy as addition, so we subtract from the blue camera signal the white sum signal of the three camera signals combined. This gives a blue-minus-white color-difference signal. Red-minus-white and green-minus-white signals are similarly obtained. Since white minus red and blue leaves green, the green-minus-white signal can be derived in the receiver, and we need to transmit only two color-difference signals, the red-minus-white and the blue-minus-white.

But how were the two color-difference signals to be transmitted? The basic idea came from RCA, namely, that a single carrier wave can carry two sets of information independently if the carrier is modulated in "phase quadrature" so that the two modulations do not interact. RCA proposed to carry the two color-difference signals on an additional subcarrier frequency, sent within the existing 6-MHz monochrome channel.

Here an idea was unknowingly used from an expired 1930 patent issued to Frank Gray of Bell Laboratories. Gray had pointed out that the energy in the spectrum of a signal generated in repetitive fashion, as a television signal generated by scanning, is not uniformly distributed over the spectrum. Instead, it is concentrated at particular regions determined by the rates of repetition; that is, there are gaps in the spectrum of the monochrome signal. If the modulated color-difference subcarrier was derived from the same timing source as the scanning, its spectrum would have similar gaps. It remained to arrange that the concentrations of energy in the color-difference spectrum should fall in the gaps of the monochrome spectrum.

Gray's contribution to the NTSC scheme was recognized in 1953, when he received the IRE Zworykin Prize. His invention is the cornerstone of compatible color television, fitting two extra signals into a channel designed for one.

# FINAL DETAILS OF THE COLOR SYSTEM

The initial demonstrations of compatible color by RCA had a basic defect. This was a fine structure of dots visible on the color images as well as on black-and-white pictures on monochrome sets. This defect was one of several reasons given by the FCC when it dismissed the RCA system in favor of the CBS incompatible one in 1950.

Bernard Loughlin, of the Hazeltine Corporation, suggested that the dots could be eliminated by two improvements. One, now used throughout the world, was the constant-luminance principle by which the color-difference signals did not affect the brightness of the received picture. The other was his proposal to apply the combined monochrome sum signal directly to the electron guns of the picture tube.

After much study and field testing, the NTSC chose 3.579545 Hz (the value used today) for the color subcarrier; this frequency fit the specifications. RCA also proposed transmission of a brief spurt of the subcarrier during the time between line scans, permitting each color receiver to synchronize exactly its own internally generated subcarrier frequency. This idea was also adopted by the NTSC after months of field testing.

# A PICTURE TUBE FOR COLOR

When RCA first demonstrated its compatible system in 1949, there was not a satisfactory single-tube display. They used three separate tubes producing images in the primary colors, these images being combined optically by mirrors. Alignment of these images to avoid color fringes was difficult to produce and maintain.

In March 1950, during the FCC hearings, RCA showed the three-tube receiver as well as the first of the shadow-mask single-tube designs. The FCC, in its decision against compatible color, also dismissed the shadow-mask tube. In this it badly misread the future, as improved forms of that tube are used in every direct-view color set today. The development of the shadow-mask tube started in September 1949, and the first model was ready for demonstration in six months. Production of an improved model started in 1951.

This tube was of great importance to the NTSC, since that body could not have conducted its field tests without an adequate color display. Principal credit for the shadow-mask color tube goes to two RCA engineers, A. C. Schroeder and H. B. Law, and to team leader E. W. Herold.

The shadow-mask tube had three electron guns, and the screen of the tube was composed of fine dots of three different fluorescent phosphors. Behind the screen was a metal mask pierced with some 200 000 tiny holes, one hole behind each group of three phosphor dots. The three electron guns at the other end of the tube were directed so that the electron beams from each were intercepted by the mask, except where the holes permitted them to pass. The alignment of the guns was such that electrons controlled by the red signal fell only on dots producing red light, and similarly for the green and blue signals and dots. The dots were so fine that the eye could not see them unaided, and the primary color images appeared to be superimposed so that the received picture seemed to be in full color.

## SUCCESS FOR THE NTSC

By the middle of 1953, after 42 months of steady work financed largely by industry, the second NTSC finally reached agreement on compatible color standards to recommend to the FCC. At the meeting of July 21, the motion to approve was seconded by Goldmark of CBS; the contest over compatibility had been ended.

The second NTSC's accomplishment was one of the great successes of technical standardization. There were 315 members of the committee and its panels, 31 of whom had served on the first NTSC. They left behind a record of 4100 pages, totaling almost a million words.

This reconstruction of a televised color picture is not commonly acknowledged to be the prodigious technical and physiological feat that it is. The televised image appears to move, but the screen presents only successive still pictures. The whole screen seems lighted at once, but actually only three tiny spots of colored light are generated. We see black, gray, white, brown, lavender, yellow, and orange. But not one of these colors is present on the screen; only red, green, and blue lights are present.

As of December 23, 1953, the FCC authorized commercial color telecasting on the standards recommended by the NTSC. The stage appeared set for a great leap forward, with RCA leading the industry, but the market was not yet ripe. The price of color receivers was yet too high, there were too few hours of color programs, and the industry lacked an all-out promotion for the new service.

It was not until 1964 that a spectacular growth of the color television industry began. During the 11 intervening years, David Sarnoff had never lost the faith. He persuaded his Board of Directors to continue to finance color work, and by the time success arrived, RCA had spent more than $100 million in research, development, manufacturing, and programming and broadcasting costs — over 100 times Sarnoff's original promise of $1 million to black-and-white television in 1935. In 1956, the industry honored Sarnoff with a gala banquet on the fiftieth anniversary of his entry into the world of wireless and radio. The President of CBS was the toastmaster.

## PROGRESS IN EUROPE
## AND JAPAN

Television was also moving in Europe. The BBC started 405-line broadcasts in 1936, as we have noted. After World War II, the French introduced an advanced monochrome system with scanning at 819 lines, 25 pictures per second, on 14-MHz channels. At an international conference in 1950, the French justified this departure on the ground that French culture could not be adequately represented with fewer picture lines.

In 1949, representatives of other European nations began meetings under the leadership of a Swiss engineer, Walter E. Gerber, to decide on postwar monochrome standards. They proposed standards with 625 lines scanned at 25 pictures per second to fit in a 7-MHz channel. The difference in repetition rates — 25 pictures per second in Europe and 30 in the U.S. — was based on the respective power frequencies, 50 and 60 Hz. The group also adopted frequency modulation for the sound accompaniment. Otherwise the American and Gerber standards were similar, and were generally adopted in most of Europe in 1950. England and France came to 625-line scanning later.

In the early 1960's, the PAL system for color originated in Germany, using much NTSC technology: a color subcarrier synchronized with the scanning rate, quadrature color modulation, and constant-luminance transmission for the color-difference signals. The PAL color system is now used in the majority of nations offering color television service. Because it transmits fewer pictures per second it is more subject to flicker; and because PAL occupies a wider channel, its detail is better than in the NTSC system.

The French — ever the individualists in television engineering — developed a system known as SECAM on an 8-MHz channel. It uses the same scanning rates as the PAL method, but it is otherwise so different that receivers designed for both PAL and SECAM (as they must be in Belgium with its Dutch- and French-oriented population) are complicated. In SECAM, the color-difference signals are transmitted sequentially, one during one line scan, the other during the next line scan. At the receiver, these color-difference signals are made to coincide in time by storing one signal until the other arrives. This process, like that used in the PAL system, makes the receiver intrinsically more expensive than NTSC receivers. Both PAL and SECAM produce excellent color reception, but PAL is clearly the favorite among the national broadcasting services.

The close ties of the Japanese electronics industry with the U.S. market led

*This Japanese equipment, made by the Nippon Electric Company, was used to transmit pictures of the Emperor's coronation from Kyoto to Tokyo in 1928.*

to the adoption of the NTSC system for the Japanese public. Variations have developed, including the Trinitron® color tube by the Sony Corporation. This uses only one gun, and the phosphor dots are replaced by vertical line segments. The shadow-mask holes are replaced by slots in the mask. The dot screen has now been largely supplanted by the line segment design in other picture tubes.

Under the sponsorship of the Japan Broadcasting Corporation, further television improvements are under study. The detail of the image has been increased by scanning at 1125 lines, 30 pictures per second, using a 20-MHz channel for the monochrome signal and 7 MHz more for the color-difference signals. The images produced compare favorably with the detail of 35-mm motion pictures. In one test in 1979, the programs were successfully received and rebroadcast from a geostationary satellite. By 1981, all the essential elements of the system had been produced in prototype form in Japan for this "high-definition television" but available channel space was lacking. Direct home reception from satellites or through cable systems appears to be practicable.

# THE ROLE OF THE
# PROFESSIONAL SOCIETY

Maintenance of the lines of technical communication was the responsibility of the AIEE, IRE, and IEEE throughout the events recounted here. Many, if not most, of the concepts and inventions mentioned were described first in the PROCEEDINGS OF THE IRE, but the role of the professional societies in television standardization was remote.

It has been the rule in the standardization activities of the AIEE, IRE, and IEEE to define methods of measurement, but to avoid entering the jungle of conflicting industrial claims for performance of equipment and systems. During the work of the two NTSC's, the IRE provided rooms for meetings and identification of members for the committee and its panels—nothing more. The RMA as a trade association (now the Electronics Industry Association) was the responsible body in television standardization in the U.S. from the beginning.

# THE IMPACT OF TV
# ON SOCIETY

The societal impact of television has been praised by many and condemned by many more. If the hoped-for partnership between television entertainment, sports, and news service on the one hand, and two-way channels of information for domestic display on the television screen develops, the impact will not be reduced. For good or ill, television offers a contact with people not possible in any other way.

The statistics of this contact are staggering. In 1983, there were nearly 600 million television sets in use throughout the world, one for every eight of the 4.8 billion men, women, and children on earth. In the U.S., 78 million of the 80 million homes have at least one television set, 68 million of these have color sets, and half have at least two sets. They are served by over 1000 commercial and educational stations. Well over one quarter of U.S. homes are connected to cable services, a figure that rises to one half in Canada. Advertisers spend more than $11 billion in sponsoring programs each year.

The impact on politics, which started with the mechanical televising of Governor Al Smith by Alexanderson in 1928, is now such that the time spent by candidates before the camera, and the impression they make on the audience, is a dominant factor in elections.

The lesson of this technical history of television is the degree to which the accomplishments of individuals and organizations had to be combined and interrelated to succeed in developing the service. In aid of this interlocutory function, maintaining the dialogue among all the participants through meetings and publications, the AIEE, IRE, and IEEE have played a unique role.

## For Further Reading

A. Abramson, *Electronic Motion Pictures*. Los Angeles, CA: Univ. California Press, 1955; reprinted by New York Times Arno Press, New York, 1974. Details of television cameras.

G. H. Brown, *And Part of Which I Was—Recollections of a Research Engineer*. Princeton, NJ: Angus Cupar, 1982. An informed, and inside view of RCA's development of compatible color television.

D. G. Fink, *Color Television Standards*. New York: McGraw-Hill, 1955. History of the development of the NTSC compatible color television standards to 1954.

——, "Perspectives on television: The role played by the two NTSC's in preparing television service for the American public," *Proc. IEEE*, vol. 64, pp. 1322–1331, Sept. 1976. Brief history of the work of the two NTSC's.

——, *Television Standards and Practice*. New York: McGraw-Hill, 1943. History of development of NTSC monochrome television standards to 1941.

——, "The forces at work behind the NTSC standards," *J. Soc. Motion Pict. Telev. Eng.*, vol. 90, pp. 498–502, June 1981. Brief account of nontechnical factors affecting the development of the NTSC monochrome and compatible-color standards.

E. W. Herold, "History and development of the color picture tube," *Proc. Soc. Inform. Display*, vol. 15, no. 4, pp. 141–149, 1974. Details of the development of the shadow-mask color picture tube, 1949–1951, with some data to 1974.

J. H. Udelson, *The Great Television Race*. University, AL: Univ. Alabama Press, 1982. An outstanding and detailed history to 1941.

The author of this chapter is deeply indebted to Prof. Udelson for his help and for the information abstracted from his book for the preparation of this chapter.

# 10

# COMPUTERS AND THE INFORMATION REVOLUTION

The Industrial Revolution, driven by steam, was underway before electric power and electrical communications appeared. The electrical arts greatly extended the influence of that revolution, and put power to use in places remote from fuel. Agriculture and manufacturing were totally transformed. But only a modest impact was felt on medicine, law, finance, government, education, and commerce. These professions are based on knowledge and analytical skill, and are more dependent on the relations between people than on mechanical or electrical power.

Now, within the short span of three decades, even these knowledge-based activities have come under the influence of a powerful new engine, the stored-program computer. This chapter explains how the sciences of electronics, logic, and linguistics have joined to provide new machines that assist the human mind. The effects on virtually every avenue of human endeavor have been so pervasive that society is now clearly in the grasp of a second great transformation, the Information Revolution.

## INFORMATION AND ITS SYMBOLS

Information is an elusive concept, defined in the 1980 *Oxford American Dictionary* as "1. facts told or heard or

*The 1645 adding machine of Blaise Pascal.*

discovered. 2. facts fed into a computer." This definition is new in that it links information and computers, but it is too narrow for our purposes. A more comprehensive definition is "meaning conveyed by symbols."

Here "meaning" refers not only to the rational content of facts and opinions, but to sensations and emotions created by our individual responses to the world around us. We are informed by eye, ear, and touch; we inform others by speech, writing, and the arts. We also convey meaning by such nonverbal unwritten messages as perfume, caresses, and the serving of a good meal.

Information is conveyed by symbols of one sort or another. A large array of symbols for reading and writing appears on the keyboard of a typewriter or computer. Such symbols are combined into words, sentences, chapters, and books. With these and other symbols the equations of mathematics, science, and engineering can be written. Music, too, has its symbols printed on the score; musicians transform them into pitch, variations in loudness, and timbre. Civilization's accumulated knowledge is passed on to future generations through storage of symbols.

Symbols represent information in its active and its latent forms. All information that can be put into symbolic form is grist for the computer, provided that the symbols have some aspect that can be quantified. Once in the computer, the symbolic information can be manipulated, interpreted, stored, retrieved, displayed, printed, and, lately, presented in speech.

In the inner workings of the computer, the symbol situation is simple. Only two marks or symbols are recognized: the numbers 0 and 1. The 0's and 1's are known as "bits," a contraction of "binary digits." All the complex outside-the-computer symbols are represented in the computer by "computer words," strings of zeros and ones.

# CODES FOR THE COMPUTER

A table that permits interchange between "human" and "machine" symbols is known as a code. An early code was that for the telegraph, devised in 1838 by Alfred Vail, assistant to Samuel F. B. Morse. He assembled a dot—dash code to match letters of the English alphabet.

The story is that he counted the letters in the local print shop, determined that *e* was most frequently used and represented it by a single dot. The letter *t* was next, to which he assigned the dash. At the end of the list, he assigned combinations of four dots and dashes to the little-used letters *q*, *x*, and *z*. His purpose was to transmit the most information in the least time.

Today, the most widely used code for operating computer printers is the American Standard Code for Information Interchange (ASCII). In its extended forms, using 8-bit combinations, the ASCII code can represent $2^8$ (256) different entries, of which 196 are assigned to different forms of letters, numerals, punctuation, and mathematical symbols. Other combinations command printer operations, including the familiar positioning operations of a typist at the keyboard. Information is most commonly fed to a computer by such a typewriter keyboard, which electrically translates the key symbols to the internal ASCII code.

# CHANNELS OF COMMUNICATION

Accuracy is the basic requirement in the transmission of information, but the rate of information transmitted and the inherent channel noise encountered in transmission over great distances impose limits on accuracy.

The speed of dot–dash transmission over the Atlantic cable (see Chapter 2) was very slow, since the advancing electric wave had to charge the cable insulation as it progressed. Open-wire telegraph lines could be operated at higher dot–dash rates, because the capacitance of the line was so much smaller. Spoken words convey information more rapidly than does the manually sent telegraph code, and higher frequencies are present. In telephone signals, the line capacitance again limits the high frequencies to about 3000 Hz and reduces accuracy in understanding speech over the telephone. Pupin's loading coils balanced out the capacitance effects, thus extending the frequency range of the lines and improving the accuracy of transmission, but at additional cost. Television signals (see Chapter 9), with their great content of information, require a wide channel (6-MHz bandwidth) at still greater cost.

The undersea cable transmits information very slowly but uses only a narrow frequency channel, the telephone transmits information at an intermediate rate with moderate bandwidth requirements, and television transmits information at a high rate and requires a very wide frequency band.

As the signal weakens in passing down the line, it is eventually masked by random currents (electrical noise). Some noise is also due to interfering signals picked up from other circuits. The masking noise poses a fundamental limit to the distance over which an accurate and intelligible message can be sent. On long lines, repeater amplifiers are spaced along the lines, regenerating the signal at points where it becomes weak. This allows transmission distance to be greatly extended.

A message can be distorted or entirely wiped out by noise and interference, or distorted so that accuracy is lost by lack of sufficient channel bandwidth. A measure of the information that can be accurately transmitted over a given channel bandwidth was proposed by R. V. L. Hartley of the Bell Telephone Laboratories in 1928. Essentially, the Hartley law says that the amount of information transmitted is determined by the time taken for the transmission multiplied by the channel bandwidth. If a given message is to be transmitted faster, then a wider bandwidth is needed. This relation was the first step into the major field of electrical science now known as information theory.

This principle had been used, if not quantified, since the earliest days of telegraphy and amateur radio. In noisy or weak signal situations the operators transmitted each word twice to ensure accurate reception. The improved accuracy was a result of increasing the time without increasing the bandwidth.

The principle is also applied in the computer world with the "parity bit." In "odd parity," this is an extra zero or one added at the end of each bit string to make the number of ones in the string an odd number. If the number of ones is originally odd, a zero is added; if the original number of ones is even, a one is added. This makes the total count of ones an odd number in both cases. The ones are counted at the receiver, and if this sum is an even number an error has been detected. More elaborate forms of this procedure add other bits and allow identification and correction of the erroneous bit. Again, the time of transmission has been lengthened to improve accuracy.

Claude Shannon entered the field in 1937, with a Master's thesis at M.I.T. in which he showed the value of two-variable (binary) algebra, using the symbols 0 and 1, in the design of electrical switching circuits. Such an algebra had been explored by Gottfried Leibniz in 1703, but he had been concerned mostly with number theory. Leibniz was followed much later by the Englishman George Boole, who in 1854 published *An Investigation Into the Laws of Thought*. Boolean algebra became the foundation for symbolic logic and for the subsequent art of computer design.

Shannon used Boolean two-variable algebra because he saw the analogy with the operation of an electrical switch as a two-state device, off and on, and he pointed out that these states could be associated with the 0/1 (true/false) symbolism of Boole.

With Shannon's application of logic to switching theory, it was possible to design switching circuits from logic statements, circuits carrying out the basic logical operations of AND, OR, NOR, and NAND (negative AND). It was also possible to reduce apparatus complexity in logic circuits, and thereby to design for minimum cost. Logic circuits are the foundation on which computers are built.

In his communication theory, Hartley assumed negligible noise in the frequency channel. In 1948, Shannon, who held a doctorate and was then working at the Bell Laboratories, expanded the communication law to obtain a numerical measure of the uncertainty with which a message arrives at its destination in the presence of noise. Perfect reception — complete information accuracy — means zero uncertainty. But noise always introduces a certain amount of uncertainty into the received signal. Shannon's law showed that for a given accuracy it is possible to interchange bandwidth and signal-to-noise ratio ($S/N$). A small $S/N$ is acceptable if bandwidth can be increased. Equally, a reduced bandwidth is possible if $S/N$ is increased, e.g., by raising transmitter power. The latter step is costly, however. Shannon placed information studies on a rigorous statistical basis, and he deserves his appellation as the father of information theory.

Shannon's work also showed that the signal form influences the accuracy of reception. Continuous (analog) signals employ all manner of wave forms that must be differentiated from noise. Transmission with pulses (digital transmission) uses rigidly prescribed waveforms, and the receiver must decide only whether a pulse is present or absent. Binary-code signals are therefore receivable with greater accuracy than are analog signals.

Space communication is based on Shannon's laws. Huge dish antennas capture large amounts of weak signals to enhance the $S/N$ ratios and thus improve the accuracy of the received information. In 1981, NASA probe Voyager II arrived in the vicinity of Saturn, some 900 million miles from Earth. Its cameras returned magnificent still pictures of Saturn and its rings. These were made possible by an extremely wide bandwidth and the use of one redundant bit for every bit carrying useful information. The transmission errors were thereby reduced to less than one bit in every 10 000 bits received. For the NASA Space Telescope mission of 1985, the accuracy will be further improved with three redundancy bits for every information bit. Only in this way can the data obtained by the telescope above the earth's atmosphere be presented in sufficient detail to satisfy the astronomers.

# BABBAGE AND
# THE COMPUTER

A computer is a machine for manipulating information. The invention of the stored-program computer is credited

*Charles Babbage, inventor of the stored-program computer.*

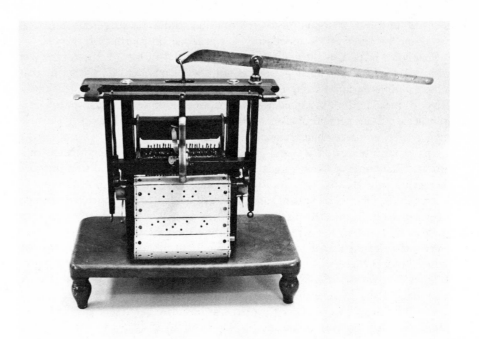

*This Jacquard loom model was carried by sales representatives in early nineteenth-century France. The belt of punched cards in front contained the weaving instructions.*

to Charles Babbage, an Englishman of extraordinary talents. Born in Devon in 1791 and educated at Cambridge University, he was a prolific inventor of calculating machines based on mechanical principles. In 1823 he obtained support from the British government to assist in development of a Difference Engine capable of computations to 30 decimal places. In 1842, after 20 years of fruitless effort to build a working model, the government support was withdrawn.

In 1828 he was appointed Lucasian Professor of Mathematics at Cambridge, a post once held by Isaac Newton. While teaching there, he conceived a machine to solve logical problems by computation, and called it his Analytical Engine. Plans for this engine embodied in mechanical form all the elements of the electronic stored-program computer of today.

Its central element, which he called a "mill," would now be called the arithmetic logic unit (ALU). It was intended to add, subtract, carry and borrow, and shift numbers for multiplication and division. It had a memory unit and a control unit for linking the mill and the memory as required. Numbers were to be entered into and extracted from the mill by punched cards, invented in 1802 by J. M. Jacquard to control the weaving of patterns in a loom.

None of this magnificent conception was reduced to practice. Babbage's mechanisms could not be made sufficiently precise or reliable to produce repeatable results. His effort might therefore be dismissed were it not for his insight into the ways in which words can enter into computation. Today he is often called the grandfather of the computer.

Babbage knew that the instructions he planned to use to direct his machine's operations, being in numerical form, could be manipulated or "computed." Properly controlled, this process would yield numbers that were new instructions. His inspiration came when he realized that such manipulations

*Grace Hopper at her desk.*

could take place within the machine. In modern terms, he knew that a computer can change its program according to its findings at particular stages in its computations.

The computer could test a number by subtraction at a particular point in the calculation to determine whether it agreed or disagreed with a number specified in an instruction. If the numbers were alike, the answer was zero; if the answer was not zero, the numbers were different. If the result were zero, computation would proceed along one line; if the subtraction showed a nonzero condition, the computer would proceed on a different line. This choice between alternatives is today known as a "branch" in the logic diagram of a computer program, and is taught in grade-school computer courses.

## BABBAGE AND
## THE COUNTESS

Babbage had a brilliant mathematician as his partner: Lady Augusta Ada, the Countess of Lovelace, born in 1815 to the poet Lord Byron. Lady Augusta was the first of many women who serve in high places in computer programming. Adele Goldstine wrote the first program for ENIAC. Grace Hopper is credited with the concept of the software "compiler" and its application to the Cobol computer language system. Jean Sammet has made major contributions to software. Mrs. Hopper and Miss Sammet are among the few female members of the U.S. National Academy of Engineering.

Lady Augusta shared with Babbage these basic concepts of computer science. The letters of words can be represented by numbers (as in the ASCII code); words can be recognized by subtracting their representative numbers; numerically

*Left: Using the Hollerith system to record and tabulate the 1890 U.S. Census.*
*Right: Grace Hopper teaching a class of programmers.*

expressed instructions can be recognized by matching them to like numbers stored in the machine; and the meaning carried by words in number form can be manipulated so as to reach logical conclusions.

With these tools, all quantifiable information is subject to computation processes by which logical conclusions can be reached. The ability of computers to deal with thought thus depends on the fact, rigorously proven by Boole, that a carefully contrived series of logical statements will always lead to a true or false conclusion.

# THE ROUTE TO THE
# ELECTRONIC COMPUTER

The grand concepts of Charles Babbage and the Countess lay unrealized for more than a century while technology caught up. Computation controlled by punched cards was one of the first steps, and these cards were joined by punched paper tape at the end of the 19th century.

A mining engineer with a flair for statistics, Herman Hollerith, was asked to mechanize the time-consuming task of keeping medical statistics. The story goes that he decided on punched cards over paper tape after watching a train conductor punch tickets. He devised a code to relate the positions of holes in a card with the 26 English capital letters, the numerals, and a group of punctuation marks. First used for mortality records in 1886 and then in the Surgeon General's office in 1889, Hollerith cards were then used for the U.S. Census of 1890. Among the companies exploiting them was Hollerith's Tabulating Machine Company. In 1924 that company became the International Business Machines Corporation — IBM.

In a modern card, there are 80 vertical columns and 12 horizontal rows. Their intersections are the locations for possible holes. One, two, or three holes in a column represent, respectively, the decimal digits, the upper-case letters of the English alphabet, and punctuation marks. The cards are prepared in a keypunch, a machine with a typewriter keyboard. Cardreaders are used to retrieve the information. In early cardreaders, the holes were detected by moving the card past an array of electrical contacts, one for each row. Reading is now done photoelectrically. In the on—off currents resulting from reading the punched card, we find a union of the symbols of printing with the binary world.

In modern computer installations, the card system has been replaced by a typewriter-like keyboard for information input and magnetic tapes and disks for storage and retrieval. Data can be recorded magnetically in extremely compact form. One 13-cm (5¼-in) two-sided "floppy" magnetic disk can hold 2 560 000 bits, which is more than sufficient to store the words of four chapters of this book.

In 1937, Howard H. Aiken, a Professor at Harvard University, started to build the Automatic Sequence Controlled Calculator (ASCC) with IBM support. It used paper tape for input and electrical relays for logical operation and data storage. It was an electromechanical form of Babbage's Analytical Engine. Completed in 1943, it ran on a three-shift basis during and after World War II, computing mathematical tables until 1948. Thus ended the electromechanical era of computation.

# COLOSSUS
# VERSUS ENIGMA—
# THE CODE WAR

During World War II, the Germans used a code machine ENIGMA. They believed it produced unbreakable code, but they were wrong. About the size of a typewriter, its replaceable code drums translated a plain-language message into coded form. After one of the ENIGMA

*The German World War II ENIGMA.*

machines was smuggled out of Poland, the British developed an electronic computer to defeat ENIGMA.

The smuggled machine was turned over to a British team including Alan Turing, a now highly respected name in computer theory. The team knew that to break the code would require reams of fast calculation, and they decided to use vacuum tubes and photoelectric tape readers. Their machine was named COLOSSUS because of its size and complexity.

COLOSSUS permitted the British cryptographers to break the ENIGMA coded messages, and thus to gain access to the flood of German tactical war information. One of the British team was later to say that Turing's work "may not have won the war, but without it we might have lost it." Such was the secrecy surrounding the COLOSSUS development that American code-breakers knew nothing of it.

# ENIAC

The first American electronic computer was built at the Moore School of Electrical Engineering, University of Pennsylvania, by J. Presper Eckert, Jr. and John W. Mauchly during the years 1937–1943. They had a contract from the U.S. Army Ordnance Department for a machine that could compute artillery ballistic data at high speed. They called their computer ENIAC, for Electrical Numerical Integrator and Calculator. For speed they used vacuum tubes, 18 000 of them, operated in the decimal system, and the program was established by a tangle of wires between plug boards. ENIAC seldom operated for half an hour before one of its thousands of tubes failed in the midst of a computation, but it was faster than any machine previously built.

Another worker in the field prior to 1940 was J. V. Atanasoff, a Physics Professor at Iowa State University. He had worked extensively with IBM punched-card sorters and calculators and saw the need for faster operation. He used the binary (0 and 1) concept and completed electronic calculating circuits and a memory using charged capacitors. Unfortunately, war work took him elsewhere, but not before a visit by Mauchly, who may have picked up a few ideas. At least, this is what the U.S. Patent Court alleged years later in giving Atanasoff priority.

The team at the Moore School was fortunate to have the services of a genius in information science, John von Neumann. Born in Hungary in 1903, he was a youthful prodigy in many fields (he contributed to the theory of implosion developed at Los Alamos which was crucial to the success of the atomic bomb). While at the Moore School, he developed the relationship between arithmetic operations, storage and retrieval of data, and the concept of the stored program. His basic pattern of the stored-program computer comprised the arithmetic logic unit, the memory, the control unit, and input/output devices. Such a design was known as a "von Neumann" computer, and is only now being displaced by newer concepts of computer architecture.

ENIAC could store only 20 10-digit decimal numbers, and it took 2 s to perform 15 operations on those numbers. This was not fast enough, so Eckert and Mauchly proceeded to design another machine, based on von

Neumann's pattern; this was the EDVAC (Electronic Discrete Variable Automatic Computer).

Von Neumann designs were also being used at the University of Illinois, where the Army Ordnance ORDVAC was under construction, followed closely by the first ILLIAC for university use. These used a memory formed by charged dots on cathode-ray tube screens.

In 1951, Eckert and Mauchly produced the first electronic computer intended for sale, the UNIVAC (Universal Automatic Computer). It used vacuum tubes, although Eckert and Mauchly were convinced that transistors must follow soon (it took seven years). The appearance of UNIVAC called for heady decisions by executives of companies offering business machines, as Eckert and Mauchly sought a manufacturer. Among the active companies were IBM and Remington–Rand. Thomas J. Watson, founder and Chief Executive of IBM, decided against the UNIVAC proposition, but Remington–Rand accepted and prepared to take over the computer market. IBM employees have described the next year as "the year of the panic," as Watson reversed his stand and put IBM in a race to overtake Remington–Rand without benefit of the details of the UNIVAC technology. How IBM succeeded serves as a reminder that technology does not always succeed on its intrinsic merit.

## TIME AND PROGRESS

In 1955, the IBM Model 650 Computer used 2000 vacuum tubes, consumed 15 kW of power, weighed nearly three tons, and could perform 50 multiplications per second; it cost about $200 000.

---

**Computers and the Information Revolution**

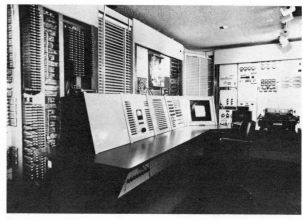

*Left: The UNIVAC incorporated a number of new technologies, among them magnetic tape input/output (at right).*
*Right: The TX-0, one of the first transistorized computers. The use of solid-state devices in computers decreased power requirements and increased reliability.*

In 1977, 22 years later, the Texas Instruments Model TI-59 hand-held calculator contained 166 500 transistors, consumed 0.2 W of power, weighed 0.3 kg (0.66 lb), and performed 250 multiplications per second. It then cost $300, which in 1983 was down to $170 at discount houses. This is a typical case in the recent history of electronic progress.

The most impressive cost reductions have come with the integrated-circuit microprocessor, a unit in which the computer functions of input/output, control, and arithmetic processing — all operations except storage — are performed by a single silicon chip costing less than five dollars. Such a chip is used in all personal computers. A spectacular example is the Timex–Sinclair microcomputer. In 1983 it could be purchased for $79.95 complete with 16 384 ($2^{14}$) bits of random access memory, keyboard, and power supply. The display is a small monochrome television set, available for another $75. Upwards of one million of these small hobby computers have been sold, each a reliable implementation of Babbage's Analytical Engine of 1834.

# COMPUTER CAPABILITIES

Computer engineers have accomplished a great deal in three decades:

*Speed*, the ability to do a great deal of work, is the strength of computers. This may represent a million-to-one advantage over corresponding human abilities. Magnitudes of work that are impossible for any manageable group of human beings become routine with a computer. Basically, the computer's speed is determined by the ability of a transistor or diode to switch on or off in a billionth of a second.

*Small size* follows almost as a necessity. Exchange of signals inside a computer requires time for signal movement. A computer signal typically travels only 15 cm (6 in) in one billionth of a second. Long wires delay signals and cause them to get out of step; delays must be short compared to the one-billionth second response time of the transistors or diodes. The wires are greatly reduced by the close integration of the elements of a chip. A memory chip that went into production in 1983 contains half a million transistors on a chip only 6 mm (0.25 in) on a side. Short paths can only be achieved in such compact integrated circuits, and producing a chip small enough for a lady's wristwatch is standard procedure. The computer industry did not waste time in noting that such chips will also perform well in large computers, at a reduced cost. In integrated circuits, smaller is cheaper.

*Accuracy* is assured by operation of transistors or diodes in on–off fashion; no intermediate values are employed.

*Endurance* is of no concern with transistors; they operate by the movement of electrons or their consorts, cannibalistic holes, as a natural process in the crystal. In normal operation there is no forced strain in the crystal material. As long as the power remains on (and there are noninterruptible power sources) and maintenance routines are observed, the data dealt with by computers are as accurate as the data input provided by the operator. When bits may be lost in transmission, as when sent over noisy telephone lines, error-correcting codes can be used. *Failures* of transistors and integrated circuits do occur, but once "infant mortality" is passed, a failure is very unlikely. Moreover, computer engineers have designed diagnostic programs which are periodically run to test the condition of all solid-state devices. If trouble is encountered, the computer will tell the operator where the difficulty lies.

The apparent *complexity* of computer equipment is largely an illusion, however real it may appear to its designers. A computer chip contains simple circuits, but appears complex because there are so many of them. The data enter as endless streams of bits, incomprehensible to ordinary mortals. Take the word "electronics," which in ASCII code is represented by 11 letters times 8 bits each, or 88 bits. It would fill two lines of print on this page. The word is lengthy but simple, since it consists of permuted strings of only two symbols, 0 and 1.

The complexity in computers has now been transferred to the process of design, grown beyond the range of human participation. Microprocessors and memory chips must be designed by a computer programmed to take one step at a time, remembering what it has done and the alternatives to the next step.

## WHAT COMPUTERS DO

The computer has now infiltrated modern business organizations and homes to such an extent that, if it were removed, most of our advanced civilization would come to a halt. A brief list of applications with which everyone has some personal contact includes:

*Direct-dial telephones.* Telephones in all major countries can now be linked by computer-controlled switches.

*Credit cards.* Use for credit checks at retail counters; billing of individual purchases.

*Banking.* Sorting of checks; printouts on monthly statements; electronic tellers.

*Supermarkets.* Scanning of universal product code on items at checkout counters to identify product and price; listing of purchase items; inventory information to the stock room.

*Social Security.* Record of account status; preparation of checks.

*Income tax.* Scanning of returns to find unusual or out-of-range entries.

*Weather forecasting.* Forecasts are based on computer analysis of atmospheric pressure patterns and movements.

*Airline operations.* Reservations and ticketing at offices, on-line worldwide.

*Industry.* Control of manufacturing processes.

*Business.* Accounting and bookkeeping.

*Hotel operations.* Reservations; fast billing at checkout.

*Medical treatment.* Tomography; diagnostic assistances; drug control; search of literature.

*Newspapers.* Typesetting direct from reporters at the scene of news events; page makup.

*Libraries.* Searches of books and magazines for information.

*Education.* Individual instruction using keyboards and displays; talking calculators for self-taught spelling and arithmetic.

*Public utilities.* Load forecasting and dispatching; system design.

*Video games.* The most popular form of home entertainment aside from television.

# PROGRAMMERS AND COMPUTER LANGUAGES

The other half of the computer profession is devoted to computer programming (software). Such practitioners are not engineers in the usual meaning of the word. They are a new breed, with talents in logic, language, orderliness, and patience with the task at hand.

Logical statements can always be written to yield yes or no answers to contrived questions. The programmer's task is to contrive the questions to which the computer can answer yes or no and yield an answer to a problem. The programs must conform to the step-by-step mentality of the computer process, and they do not directly provide the bit strings the computer uses internally. Rather, they use ordinary words or abbreviations and symbols as computer language, with meanings assigned and stored in the computer memory to be recognized and recovered as the program proceeds.

The computer recognizes these words and symbols through lists and instructions fed into the computer as software. Other words and abbreviations for such standard languages as Basic (Beginners All-Purpose Symbolic Instruction Code) are built into the computer during manufacture as read-only-memory (ROM). These circuits, when they encounter a recognizable program word, issue the appropriate bit strings to instruct the machine to perform the required operations.

Other standard languages are available and may be placed in computer memory from prepared magnetic tapes or disks. These languages include Fortran, suited to problems in engineering and science, and Cobol, to handle operations in business offices. There are dozens of others—Algol, APL, C, Pascal, and PL/1, plus specialized programs for various fields.

A most ambitious language under development is ADA, sponsored by the U.S. Department of Defense, and named for the Lady Ada, the Countess of Lovelace. Still incomplete, its purpose is to bring order into the Babel of languages now in use in armed forces computers—but it may also prove too complex to be cost effective.

# THE MICROPROCESSOR

The more computers mature, the smaller they become. In 1969 at Intel Corporation, M. E. Hoff dealt with a Japanese customer's desires for customized calculator integrated circuits. The work resulted in what he described as "a general-purpose computer programmed to be a calculator." It included four chips in a set as the 4004 microprocessor. It was followed by the 8008, an 8-bit microprocessor, the first in what has now become a great variety of microcomputers to be used where the problem arises rather than bringing the problem to a central computer. By 1982, 32-bit microprocessors were in production.

In 1973, a patent was issued to G. W. Boone, also of Intel, for a "Computing System CPU" (central processing unit) that placed on one chip all of the elements of a computer except its main memory. The power of fast computation has become available through the microprocessor for virtually every avenue of human endeavor. The "personal" microprocessor computer is now in millions of homes, and the end is nowhere in sight.

# SIGNAL CONVERSION

The workaday world does not provide digital signals. Its physical and chemical quantities vary continuously in time, described as analog variations. Electrical engineers have developed circuits that measure those analog quantities, translating their values at specified sampling intervals into strings of bits. These are analog-to-digital (A-to-D) converters. The translated signals may be entered in digital form into computers for manipulation. The reverse process of generating a continuously varying electric current from a bit string is done with digital-to-analog (D-to-A) circuits.

A widespread application of such converters, and of the microprocessor as well, is in the engine controls of every new motor car sold in the U.S. since 1981. These employ computer control of the engine to meet governmental standards on gas consumption and air pollution. All the fine balances among

*Left: The Intel 4004 microprocessor. Right: A computer on a chip: the Intel 1702 is erasable and programmable, and contains memory space as well.*

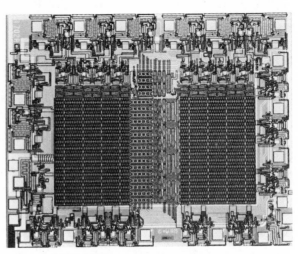

pressures, temperature, engine revolutions, throttle setting, and the like are set by a microprocessor.

The computer input is derived from signals originally in analog form, converted to digital form, and fed to the computer chip. The computed outputs are converted back into analog signals that control the carburetor mixture (or the fuel injection pump) and the spark timing to make the engine run well.

Industry uses similar A-to-D and D-to-A chips by the thousands to operate controls of chemical plants, oil refineries, power plants, and the like.

## GAMES PEOPLE PLAY

Electronic games, first exemplified by computers capable of playing championship chess or checkers, have now proliferated into a multitude of special games played in arcades or at home; most are of the destroy-or-be destroyed type.

In 1950, Arthur Samuel of IBM devised a checker-playing program of championship quality. The program was perfected by having one IBM Model 7094 computer play another until the games ended in ties! A program to play backgammon, composed at Carnegie-Mellon University in 1979, won the world championship by taking four out of five games from a disconcerted Italian, Luigi Villa of Milan.

The computer–TV screen game was started in 1962 by an M.I.T. graduate student, Steve Russell, with a game called "Space War."® Two other students, Nolan Bushnell at the University of Utah and Bill Pitts at Stanford University, independently set out to commercialize the idea. More successful was the electronic version of table-tennis, "Pong,"® produced in 1972 by Atari, a company founded by Bushnell to carry his ideas into production. "Pong" was an important sales item for Christmas in 1975. In late 1976, the Atari Company was sold at a sensational price, and Bushnell "retired" to consulting on video games and related projects.

In 1966, Ralph Baer of Sanders Associates, Manchester, NH, began work on TV games. A multigame model was sold to Magnavox in 1969 and produced in 1972, selling nearly 100 000 games as "Odyssey"® that year. Coleco was an early user of a new single-chip design for games and captured a major share of the video game business in 1976. Other companies have tried the market and died almost as fast as new companies have been born.

All this activity among young players is an excellent lead-in to the use of more sophisticated computers in the home, on college campuses, and in small businesses.

## COMPUTERS AND THE HUMAN MIND

Computer successes in game playing and theorem proving are readily explained. First, the subject matter involves measurable symbols that can be converted into bit strings and subjected to logical manipulation. Second, the programs have been designed to imitate shortcuts taken by the human brain (heuristics). Imitation of the computer-like operations of the brain by computers is no longer in question, but other forms of human thought processes seem far beyond the range of computer science, at least for many decades.

The difficulty is that many human reactions, such as the appreciation of poetry and humor, have thus far eluded symbolic measurement. The vast memory store required for the elements of these subtle forms of communication, and the means by which the brain cells call up these elements, remain far beyond the programmers' art.

We cannot guess how long it will take to achieve the density of circuits and rapid access to stored data that will match the brain in these noncomputational functions, although progress in duplicating speech is coming fast. The pace of the last three decades has been truly impressive, and new technology and theories of information organization can be expected to open new doors before the 20th century passes.

It will indeed be an eventful day when a computer, without prompting, prints the message *Cogito ergo sum*.

## *For Further Reading*

J. Campbell, *Grammatical Man — Information, Entropy, Language and Life.*   New York: Simon and Schuster, 1982.

C. Evans, *The Micro Millennium.*   New York: Viking, 1979. A look at the future of the Information Age.

D. G. Fink, *Computers and the Human Mind — An Introduction to Artificial Intelligence.*   New York: Doubleday Anchor, 1966.

D. Hanson, *The New Alchemists — Silicon Valley and the Microelectronics Revolution.*   Boston, MA: Little, Brown, 1982.

M. E. Jones, W. C. Holton, and R. Stratton, "Semiconductors: The key to computational plenty," *Proc. IEEE,* vol. 70, pp. 1380–1409, Dec. 1982.

T. Kidder, "The Microkids and the Hardy Boys," *IEEE Spectrum,* p. 48, Sept. 1981.

_____, *The Soul of a New Machine.*   Boston, MA: Little, Brown, 1981. Historical account of the development of a 32-bit minicomputer at the Data General Corporation.

J. W. Mauchley, "On the trials of inventing Eniac," *IEEE Spectrum,* p. 70, Apr. 1975.

C. E. Shannon, "A mathematical theory of information," *Bell Syst. Tech. J.,* vol. 27, pp. 379–423, July 1948, and pp. 623–656, Oct. 1948.

Special Issue on the Mechanization of Work, *Sci. Amer.,* vol. 247, Sept. 1982.

N. Stern, "From Eniac to Univac," *IEEE Spectrum,* p. 61, Dec. 1981.

H. D. Toong and A. Gupta, "Personal computers," *Sci. Amer.,* vol. 247, p. 186, Dec. 1982.

D. L. Waltz, "Artificial intelligence," *Sci. Amer.,* vol. 247, Oct. 1982.

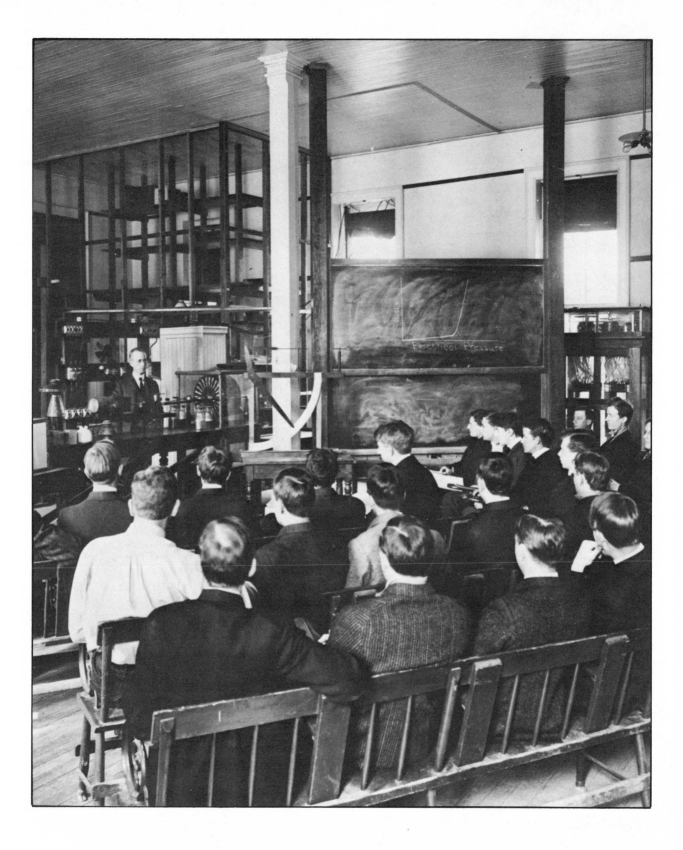

# 11

# YOUNG ENGINEERS AND THEIR ELDERS

*"There must be something in electricity, though what it is I would not venture to say, which attracts the younger and more vigorous members of our race to its study. Perchance it may be that in this mysterious force, there exists some lingering traces of the long sought for "fountain of youth," but be what it may, I find in the fact that some comparatively young men have been able to do so much for the world's weal in a special science, a bright promise of what they may be able to accomplish before their tasks are done."*

AIEE Presidential Address by Prof. E. J. Houston, May 15, 1894.

The education of the young has always been a responsibility of the elders, and young engineers have rarely been consulted concerning their mode of education during school. Although today we do have class questionnaires for instant feedback, they mainly concern the quality of teaching. The lack of student input on curriculum is not surprising, since it takes a working lifetime of 40 years to form a reasonable opinion. For instance, the electrical graduates of 1930, steeped in 60-Hz power, can only recently have seen that their deficiencies in the physical sciences posed difficulties as the new fields of radio and electronics arose.

The engineering educators, many being young men themselves, over the years have hypothesized and tested with their willing laboratory guinea pigs. By trial,

*Left: William A. Anthony, Cornell Professor of Physics, whose students included Harris J. Ryan, Dugald Jackson, and E. L. Nichols.*
*Right: Dugald Jackson built a strong electrical engineering course at Wisconsin before becoming the foundation of the M.I.T. program of the twentieth century.*

they have produced a successful system that, when well implemented, meets the students' educational needs for a lifetime — albeit with a necessity for continued learning as our dynamic field moves on. This chapter, then, is the story of those tests and trials.

# LIBERAL AND PRACTICAL EDUCATION

Technical education was alien to the universities of the young United States until Congress authorized President Washington to establish a school for engineers at West Point, NY, in 1794, and gave it more formal approval as the United States Military Academy in 1802. Under Colonel Sylvanus Thayer, after 1817, this school developed similarities to the French engineering schools of the day, with annual classes, a curriculum based on civil engineering, and a high standard of achievement.

Rensselaer Institute at Troy, NY, gave its first degrees in civil engineering in 1835 to a class of four. Its name of Rensselaer Polytechnic Institute became official in 1861, and in 1862 it featured a four-year curriculum with parallel sequences of humanities, mathematics, physical science, and technical subjects.

At that time there were about a dozen other engineering schools. The larger universities, including Harvard and Yale, were oriented to the education of children of the upper classes for careers in law, medicine, religion, and the arts. Engineering students admitted to these universities often had less than the normal high-school work in the classics, and as a result they did not always enjoy full stature on those campuses.

The lack of educational opportunities for children of the working classes was noted by many, and led to the Morrill Land Grant Act of 1862. Senator Morrill introduced his bill in 1857 and it was defeated; it passed in 1859, but was vetoed by President Buchanan; it was finally passed and signed by President Lincoln in 1862 in the absence of the southern delegations, to whom "working classes" had meant slaves. The Act gave to the various states grants of federal land, by the sale of which the states were to establish colleges featuring "agriculture and the mechanic arts, for the liberal and practical education of the industrial classes." Within ten years 70 engineering schools had been established, including many of the large institutions of today. C. F. Scott (Ohio State '85) and B. J. Arnold (Nebraska '97) were the first graduates of land-grant schools to reach the presidency of the AIEE in 1902–1903 and 1903–1904, respectively. There have been many more.

As industry developed after the U.S. Civil War, there arose a need for technical education in fields other than buildings, bridges, roads, canals, and railroads. Steam was the paramount prime mover, and mechanical engineering developed as a separate field. Yale, M.I.T., Worcester, and Stevens had early curricula in that area.

The curricula in the early engineering schools were modeled on the disciplines prevalent in the arts: basic science, mathematics, languages and social studies, and engineering knowledge and techniques. Opinion soon split on the relative importance of the practical and theoretical aspects of engineering teaching. This led to schools with opposite orientations, some employing mathematical and scientific approaches (M.I.T., Stevens, and Cornell) and others (Worcester, Rose Polytechnic, and Georgia Institute of Technology) stressing shop work and producing graduates trained as shop foremen.

*Prof. Charles Cross's laboratory at M.I.T. in the 1870's, showing some acoustic apparatus, a large static electricity generator (the disk at right rear), and a battery of electrical cells (on the floor, right front).*

# THE ROOTS
# OF ELECTRICAL
# ENGINEERING

Electrical engineering usually began as an option in the physics department. Thus, its antecedents were quite different from those of civil and mechanical engineering, which developed under the pressures of industrial demand. The first such optional course was organized at M.I.T. in 1882 by Prof. Charles R. Cross, head of the Department of Physics and one of the AIEE founders. The next year, an electrical course was introduced in the Department of Physics at Cornell by Prof. William A. Anthony (AIEE President, 1890–1891). By 1890, there were many such courses in physics departments; this parentage gave electrical engineering its roots in the sciences.

Prof. D. C. Jackson (AIEE President, 1910–1911), the first Professor of electrical engineering at the University of Wisconsin, wrote in 1934 that "our modes of thought, our units of measurement, even our processes of education sprang from the science of physics (fortified by mathematics) and from physicists. The precise measurements and controlled experiments introduced into our field from the field of physics gave a tremendous impetus to rational and accurate engineering calculations and also left a scientific impress on electrical engineering teaching . . . Civil engineering, mining engineering, and mechanical engineering, as the older major branches, arising . . . before the sciences were adequately expanded to provide a rational foundation for engineering practice, still maintain a large proportion of empiricisms in their college curricula."

The 1882 electrical engineering curriculum at M.I.T. showed physics courses throughout all four years, as the electricity was taught in those courses. The courses emphasized mathematical concepts, and advanced mathematics appeared in the final year. A year of shop work was included, specified as wood and metal turning and carpentry. Surveying was not included, although many engineering schools thought it necessary for a well-rounded engineer. History and literature courses appeared over four years, and a year and a half of German language was required, with advanced German recommended. This was important since much of the scanty literature of electrical engineering was in German, and that country was the place to go for graduate study in physics.

In 1886, the first department of electrical engineering appeared at the University of Missouri, but was hampered in development because of faculty losses. The University of Wisconsin organized its department in 1891 with D. C. Jackson as its head. He came from the Edison General Electric Company in Chicago, but only after he was assured that the department would be on an equal footing with other academic departments.

# THE AIEE AND
# EDUCATION

The AIEE heard its first papers on education in 1892. Prof. R. B. Owens of the University of Nebraska showed that

*Left: The Cornell program was first housed in the Physical and Chemical Laboratory, which was renamed Franklin Hall — in honor of "The First American Electrician" — at the end of the 1880's.*

*Below: In this 1885 engraving, Cornell students are shown testing a dynamo on a Brackett cradle dynamometer, an instrument invented by Cyrus F. Brackett of Princeton.*

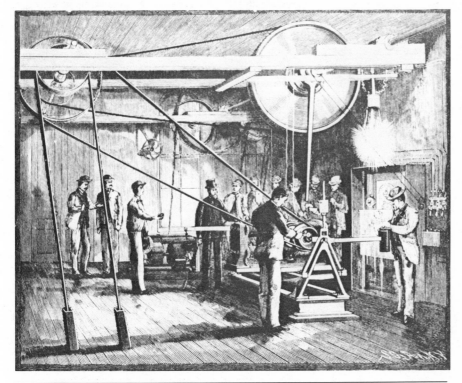

**Young Engineers and Their Elders**

scientific training was foremost in his thinking when he called for separate schools for teaching mechanic arts. He wrote, "The universities will find it no longer necessary to devote their energies to such elementary work, but will reserve their whole force for the higher instruction for which they are originally designed."

For the electrical curriculum he thought that "a good course in modern geometry, differential and integral calculus, say according to Williamson, and differential equations say, Forsyth or Craig, would be sufficient to pursue with advantage nearly any courses in electrical engineering . . . But to attempt to analyze the action of alternating current apparatus without the use of differential equations is no very easy task . . . and when reading Maxwell, it becomes convenient to have quaternions or spherical harmonics, they can be studied in connection with such readings." Although shop work and electrical installation courses were listed in the curriculum, they were subordinate to English literature, German, and Romance languages.

The work in the laboratory was developed around large machines — a 500-light 1000-V dc generator, two 15-kW 120-V generators, and two 25-light 10-A arc machines. All these were belt-driven by steam engines. An assortment of storage batteries, Prony brakes, and instruments was available. The description makes the laboratory appear a Spartan place — but this was 1892.

## CONFLICTING
## INDUSTRY VIEWS

By 1903, faculties at many schools did not agree with Owen's suggestions for curriculum. In that year, the AIEE held a joint summer meeting with the newly formed Society for the Promotion of Engineering Education (since 1946 the American Society for Engineering Education, ASEE). The papers given by industrialists suggested that engineering students should be taught "engineering fundamentals," but no mention was made of just what those fundamentals were to be. They were also to study trade catalogues and manufacturer's drawings, and to learn by visits to electric light and power plants.

There was disagreement on the value of design courses. One speaker urged their inclusion in the senior year. These would cover the materials of electrical engineering as well as study and calculation of dynamo-electric machinery and transformers. But work at the drawing board was to be at a minimum.

Steinmetz in his presidential address in 1902 had stated the opposite view: "The considerations on which designs are based in the engineering departments of the manufacturing companies, and especially the very great extent to which judgement enters into the work of the designing engineer, makes the successful teaching of designing impossible to the college." He had gone on to urge more use of analysis of existing designs. Despite Steinmetz' advice, the teaching of design courses continued into the 1930's.

In the first two decades of the 20th century, there was no ready source for shop supervisors, and many employers expected their recently graduated electrical engineers to be sufficiently trained in shop skills to take over as foremen. This was a responsibility they were not trained to fill. Complaints from such employers were registered with the schools as in their view, the graduates were not sufficiently "practical."

However, the larger electrical manufacturers developed programs in their plants for inducting college graduates into the engineering world. These programs, in which the new engineers could move to various jobs and departments of the company, were highly regarded by the graduates as offering a breadth of experience not available in more specialized companies. Thus was supplied some of the "practical" training not available on the campuses.

*The student laboratory in the Rogers Building at M.I.T. in the 1870's.*

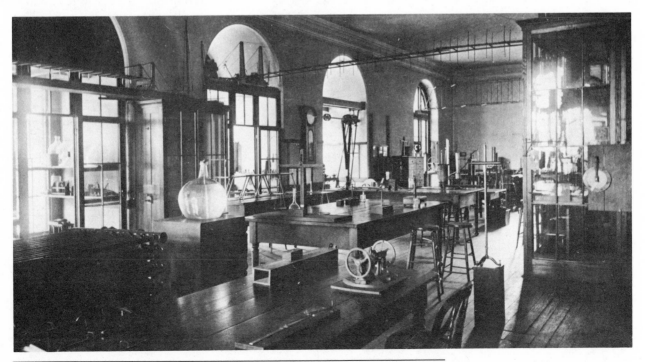

# EDUCATION
# AT A STANDSTILL

Complaints about the practicality of graduates found acceptance, particularly in schools in which electrical engineering was strongly affiliated with the more practical programs in mechanical engineering. These curricula usually lacked the science roots offered by association with a good physics department. Because many educators in the various engineering disciplines heeded the signals from industry, shop work with carbon steel tools continued after the arrival of the high-speed cutter, blacksmithing and patternmaking were taught after welded frames replaced castings, and testing of reciprocating steam engines continued after the coming of the turbine. Such out-of-date courses persisted into the 1940's.

Dc circuits, ac circuits, a junior year of dc machines including armature windings, a senior year covering synchronous machines, induction motors, transformers, and a course in design of transmission lines was the usual academic fare. Mechanical and electrical design was included despite the advice of Steinmetz. Electric railways and illumination were often available as electives. Such specialization was not so strong among the eastern schools, although even there the physics and mathematics content of the electrical curriculum was reduced.

Typical of the times were laboratory courses in which the objective was to learn the AIEE standard tests for machinery with emphasis on performance details. Vector diagrams were used but not the equivalent circuits of the machines. One professor stated that equivalent circuits and Thevenin's theorem had no place in power courses.

Materials courses, taught by civil engineering professors, covered timber, concrete, and mild steel. Not mentioned or well understood were conducting materials, magnetic materials, or dielectrics. The mechanical engineering professors taught little thermodynamics to young electrical engineers — "It was too

*This apparatus — tubs of water, battery of cells, and galvanometer — was used at M.I.T. for testing the insulation of wire. It could test to over 100 000 M$\Omega$. (The galvanometer is in the box behind the battery, hooded to facilitate reading.)*

difficult." Instead, they described boilers, steam engines, and pumps. Differential equations, thought necessary by Owens in 1892, were relegated to graduate study.

Engineering education was not much affected by World War I, despite the advances made in radio and aviation. The annual report of the AIEE Education Committee in 1921 states, "The methods of giving instruction and the training given to engineering students now do not differ greatly from what obtained before the war."

Engineering teaching in the years from 1900 to 1935 was almost static; professors largely held bachelor's degrees and had some years of practical experience, which was viewed as more important than theoretical or mathematical proficiency. Very few teachers held Master's degrees, and far fewer had doctorates. Many teachers assumed that a student would need no knowledge beyond what the teacher had employed in his career. There was no thought of preparing the students for change in the field, although there was a radio receiver in every professor's home.

As a case in point, Steinmetz had masterfully presented the case for the use of complex algebra and *j* in circuit analysis before the AIEE in 1893. But this analytic technique, which outmoded the cumbersome vector diagram in ac circuit calculations, had not been fully adopted on the campus by 1925. In fact, it was necessary to teach complex algebra from mimeographed notes, the subject having not yet appeared in some textbooks.

Engineering education was not well financed. Money was diverted to agriculture in the land-grant schools, as research in that field affected people directly through increases in farm productivity. Research in engineering was not considered necessary, for as one academic department head wrote, "had it not all been done by the inventors?" Dr. F. E. Terman (IRE President, 1941) reported on electrical research productivity in 1920–1925. He found an average of nine technical papers per year from university authors published in the AIEE TRANSACTIONS, but three-quarters of those papers came from only five institu-

*These measuring instruments were all used in the M.I.T. electrical engineering laboratory in the 1880's and 1890's.*

*Left: Harris J. Ryan, who left Cornell in 1905 to head the electrical engineering program at Stanford University.*
*Right: Frederick E. Terman, whose perseverance and students built California's Silicon Valley.*

tions. In the PROCEEDINGS OF THE IRE, over half of the 30 papers from educational sources published in those five years were submitted by authors in physics departments.

In the stagnant years from 1900 to 1935, there were pressures on engineering educators from employers and alumni to increase the teaching of industrial practice. During that time, electrical education lost much of its cultural heritage, both in technical areas and the liberal arts.

Discouraged by boring work on the drafting table and in the shop and by heavy class schedules, good students became disenchanted with engineering and transferred to a new and competitive field, engineering physics. This field arose because progressive industries saw the need for more science and mathematics as preparation for work in the industrial research field that began to flourish in 1925. Many electrical graduates undertook graduate study in physics.

Thus the three decades preceding 1935 were a particularly unfortunate time for electrical engineering education, for it was in those same years that great advances were made in its parent science, physics. Planck's radiation law, Bohr's quantum theory, wave mechanics, and Einstein's work offered answers to many questions. But electrical engineering education was not aware of their significance. Even radio, well advanced by 1930, was overlooked by electrical educators, except in a few schools where dynamic young men were able to teach courses in radio and communication.

## DEVELOPMENT OF
## GRADUATE STUDY

The complexities of radio and electronics attracted some students to a fifth year of study for a Master's degree. The large manufacturers did not recognize the value of a Master's degree until 1929, when the time-honored salary of $100 per month for a starting engineer was raised to $150, with 10 percent added for a Master's degree. The power field, too, had problems worthy of a year of advanced study, but the public utilities, desiring management skills as much as technical ones, were slow to see the value of an extra year on campus.

The value of graduate work and on-campus research programs was dramatically demonstrated just before World War II, when new electronics-oriented corporations, financed with local venture capital, began to appear around M. I. T. and west of Boston. These were owned or managed by doctorate graduates from M. I. T. or Harvard who had developed ideas in thesis work and were gambling on a market for their products. Another very significant electronics nursery appeared just south of Palo Alto, CA. This area is now known as "Silicon Valley" for its abundance of semiconductor and computer manufacturers. This industrial development resulted from Dr. F. E. Terman's encouragement of electronics research at Stanford University. For example, the Hewlett-Packard Company, now a multibillion dollar operation, started with a $500 fellowship to David Packard in 1937. He and William Hewlett developed the *RC* oscillator as their first product, and manufacture started in a garage and Flora Hewlett's kitchen. Many other companies in the instrumentation and computer fields have similar backgrounds and histories.

# EFFECTS OF
# WORLD WAR II

World War II had a tremendous impact on engineering education. Many schools participated directly in Navy and Army programs, or in the Engineering, Science and Management War Training program for producing technical manpower. After the war, the engineering colleges faced a flood of young people financed by the G.I. bill, children of workers who were aware that college work was necessary for economic betterment.

*Terman (center) with two of his better-known students, William Hewlett (right) and David Packard.*

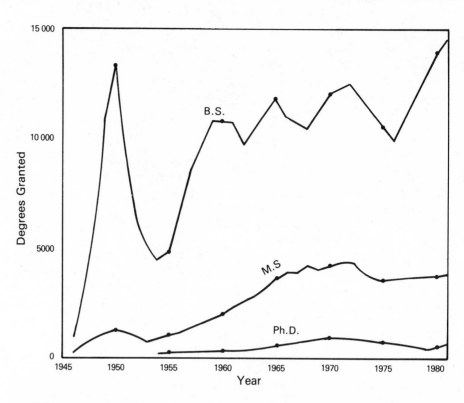

*Fig. 11.1: Electrical engineering degrees granted from 1946 to 1981.*

Of these, many had seen or operated the electronic marvels produced for the war effort. They were a new generation intrigued by the mysteries of the electron. Electrical engineering became the largest engineering branch as enrollments soared (Fig. 11.1). Electrical and computer engineering enrollment at the Bachelor's level exceeded 120 000 in 1982.

Research sponsored by industry and government created a significant increase in graduate enrollment as the students sought the additional study they needed to become contributing researchers or teachers. By the early 1970's, about 30 percent of the Bachelor's level students went on to receive a Master's degree, and 8 percent followed to take the doctorate. These percentages have declined in the 1980's because of the attractiveness of industry offers and because of the reduced government support of graduate study and research.

A Bachelor's degree thesis was common before about 1930, but this requirement gave way to the pressures created by too many students and by the lack of equipment and ideas for study. After the war the Master's thesis also tended to disappear, again under pressure of numbers. By then, the Master's work had often become simply a fifth year of academic study.

The increased numbers of students seeking admission at the undergraduate level allowed the engineering schools to be more selective, and the abilities of the students were higher than in the prewar years. Shop work and engineering drawing were replaced by computer programming as the modern engineer's language. In the electrical curricula there was more emphasis on circuit theory, mathematics, physics, and electronics, and a reduction in the study of power apparatus. It was said that students had by then learned that $\omega = 2\pi f$ was a variable, whereas in prewar classes many had thought it a constant equal to 377.

Most of this curricular change was made necessary by the failure of World War II engineers as researchers in the laboratories exploring the new fields opened by the war. Many of the earlier graduates, whose education had centered on 60-Hz power equipment and design, did not have the mathematical and research training needed in the new fields of electromagnetic waves, pulses, radar, control systems, and electronic instrumentation. Physicists had to play the leading roles in wartime research.

## AND THE FIELD
## MOVES ON

Further advances were ahead for which the electrical educators had to prepare their students. By 1950, solid-state materials and devices were in view, requiring the teaching of conduction processes that were based on the physical theory of the previous decades. Today's electrical engineering comprises circuit theory, field theory, energy transformation, feedback and control, and information processing. Compared to pre-World War II curricula, its study relies much more heavily on principles, and a device or machine is studied not in isolation, but with its connected system. The physical theory ignored in the 1900–1930 decades must now be studied as fundamental to most of our electrical theory.

The control area is an example of a new field. Although servomechanisms were around as early as 1934, wartime applications made control important. Computer science came from switching theory and from logic circuits that used diodes and transistors for their high response speeds. As computer software has become more important, embodying the properties of a language, some computer departments have separated from electrical engineering.

So has the electrical field turned one sharp corner after another. If the fundamentals — the physical and mathematical tools — were not properly covered, it was difficult for the practitioners to negotiate the corners into new careers.

## MEASURING
## ENGINEERING
## EDUCATION

The engineering profession, through the Society for the Promotion of Engineering Education, several times conducted appraisals of its education programs. In 1907–1918, such an effort, financed by the Carnegie Foundation, produced a report that urged the development of the students' intellectual abilities rather than the teaching of facts and techniques. In the period from 1923 to 1929, a committee under the chairmanship of W. E. Wickenden, later President of Case Institute, surveyed all aspects of the engineering education effort. It found a great diversity in levels of engineering education, and suggested these be measured and controlled by the accreditation of engineering curricula. This proposal led to the founding of the Engineering

Council for Professional Development (ECPD) to undertake that accreditation through on-campus visits by teams of educators and engineers.

About 1940, with its work interrupted by the war, a committee under H. P. Hammond of Pennsylvania State University prepared an SPEE report that urged an increase in the content of humanities and social studies. This was a call for a return to some of the earlier classical influence in programs for electrical engineering, such areas having been lost under the pressure to teach specialized technical material.

In 1946, the SPEE changed its name to the American Society for Engineering Education (ASEE), and later continued its program of educational improvement by another survey, this time under Dr. L. E. Grinter, now of the University of Florida. This report (in 1955) again urged increased efforts in the social-humanistic areas to give engineers more understanding in dealing with people. A list of "engineering sciences" was proposed for all engineers' study and included mechanics of solids, fluid mechanics, thermodynamics, heat and mass transfer, electrical theory, and the nature and properties of materials. The committee's recommendation on engineering science had significant impact.

Another survey of engineering education was undertaken in 1968 under the direction of Dr. Eric Walker, then President of Pennsylvania State University. That report asked schools to be aware of the increasing demands on the engineers' social consciousness, noting that engineers' decisions were more often affecting people as well as things.

These surveys were all effective in moving engineering education forward, each contributing in its own time. The 1940 report increased the attention given to social-humanistic study; at the time, that area of study had been severely reduced. Particularly significant was the Grinter report of 1955, which reached the schools at an especially propitious time — during the post-war realignment. New faculty, replacements for the retirees of the war years, largely held doctor's degrees and were eager to try their skills in the incorporation of more science into their courses.

Accreditation of curricula is now carried out by teams responsible to the Accreditation Board for Engineering and Technology (ABET), successor to the ECPD. The IEEE has representatives on the governing board of ABET and assembles teams to conduct school visits for electrical engineering, electronics, and computer science and engineering. The visiting teams appraise faculty competence, student selection practices, curriculum, and the resources of the institution under inspection. Curricular content is studied in mathematics, science, engineering sciences, engineering design, laboratory, computer usage, and social science and the humanities. In 1982, there were 239 electrical and computer curricula in the United States on the accredited list.

# COMMUNITY COLLEGES AND TECHNICAL INSTITUTES

Many states have two-year community colleges. These schools seek to serve their communities through post–high school technical and service-oriented education, and to provide the first two years of four-year college programs. Students in the latter area may ultimately transfer to four-year schools. Engineering colleges usually have good relations with these institutions and endeavor to see that technical work taken at the community college is at a level suitable for transfer into the college engineering curriculum.

There are also two-year technical institutes and two and four-year technology programs, specifically designed to supply hands-on programs for students not interested in the more theoretical and abstract work of the engineering colleges. The graduates are sought as supervisors and engineering assistants in industry. The degree given after two years is Engineering Associate. The four-year program leads to the Bachelor of Technology degree.

In the fall of 1982, in the two-year courses in the U.S. there were 24 200 enrolled in electronics, 5200 in electrical technology, and 6700 in computer work. The corresponding figures for the four-year B.T. programs were 10 800, 5100, and 900.

# CONTINUING
# EDUCATION

Contrary to an earlier view of engineering as a field calling only for application and improvement of existing methods and devices, it is now recognized that engineers must be able to move into new fields and work with devices unknown at the time of their basic university education. This calls for a continuation of the educational process throughout an engineer's working lifetime. Such continued education of engineering graduates has always been considered a responsibility of the colleges, but it is one not easily handled. The responsibility comes from the extension activities that are part of the land-grant heritage of many engineering schools. Evening graduate courses on campus serve well in schools located in metropolitan areas, but the diversity of student interests makes it difficult to gather enough students from areas beyond commuting distance.

Companies in isolated locations may provide facilities for their employees for telephone conferences, electronic blackboards, or videotape presentations. These new communications techniques are suitable for services once supplied by a college at a distance, and now we have the telecollege. Twenty-four university programs have successfully employed live television over microwave relay, with over 45 000 engineers currently enrolled where they work. Industry or government subsidy is usually required for financing these projects.

Another avenue to improve skills on the job is the intensive short course on a specific new technical topic. Although expensive to the individual, they are a means of providing new skills or giving an entry to a new field. They are offered by various engineering colleges in major industrial cities in their service areas.

The IEEE Educational Activities Board has undertaken to provide more education for engineering graduates, directed to improvement of on-the-job performance. They have developed a number of correspondence courses using audio casettes. The IEEE has also employed satellite-transmitted special lectures to remote groups, and the use of videotape is increasing.

# CONTRIBUTIONS OF
# ENGINEERING FACULTY

With the current underfunding of education on a national scale, it is estimated that engineering faculties are short of personnel by 10 percent at all ranks and that equipment shortages approximate one billion dollars. Researchers may work with better equipment in industrial laboratories, an attraction that lures faculty away from the campus. Industry bidding for graduate students, in competition with universities, is cutting into the number available for the teaching of undergraduate and graduate students.

The professors have always constituted a group interested in students, in education, and in the profession. As can be seen in the Appendixes, professors have often served as Presidents of one of the Institutes; a great many others from academic life have been IEEE directors and committee members. One of their greatest contributions to the profession has been their textbooks, by which our professional knowledge has been refined and carried forward.

Dr. W. L. Everitt (IRE President, 1945), then of Ohio State University, published *Communication Engineering* in 1932 and established the field. He identified the elements of circuit theory essential to signal transmission over a wide range of frequencies. The book stressed fundamental principles rather than applications or apparatus, departing from previous textbook styles. At about the same time, Dr. F. E. Terman published *Radio Engineering,* which gave the same fundamental treatment to vacuum-tube usage in the radio frequency field. Both of these books showed that the new communications area required fundamentals in circuit theory and mathematics not previously considered in courses covering the limited range of power frequencies and applications. These textbooks remained standard approaches until new needs emerged after World War II.

# BROADENING OF
# THE FIELD

In the stagnant decades, it was common to hear a professor state that students should not graduate without taking a particular course or "seeing" certain apparatus. This reflected the narrow view of the objectives of engineering education, considering it only as the acquisition of facts. The training of students' minds in the solution of problems, using engineering methods and the principles of science, is the aim of engineering education today. This facility is joined to development of familiarity with liberal culture to ensure a useful and satisfying life. This makes engineering programs a good general education, suited to follow regardless of the corners around which the electron may lead.

# For Further Reading

L. V. Baldin and K. S. Down, "Educational technology in engineering," in *National Academy of Engineering.* Washington, DC: Nat. Acad. Press, 1981.

L. P. Grayson, "A brief history of engineering education in the United States," *Eng. Educ.,* vol. 68, p. 246, 1977.

L. E. Grinter, "Report on evaluation of engineering education (1952–55)," *Amer. Soc. Eng. Educ.,* vol. 49, p. 36, 1955.

D. C. Jackson, "The evolution of electrical engineering," *Elec. Eng.,* vol. 53, p. 770, 1934.

J. F. Mason, "The secret society that never was," *IEEE Spectrum,* p. 55, Sept. 1979.

R. B. Owens, "Electro-technical education," *AIEE Trans.,* vol. 9, p. 462, 1892.

H. L. Plants and C. A. Arents, "History of engineering education in the land-grant movement," in *Proc. Amer. Ass. Land-Grant Colleges and State Universities,* vol. 2, 1961.

R. Rosenberg, "American physics and the origins of electrical engineering," *Phys. Today,* Oct. 1983.

——, "The first years of American electrical engineering education," *IEEE Spectrum,* to be published.

F. E. Terman, "A brief history of electrical engineering education," *Proc. IEEE,* vol. 64, p. 1399, 1976.

——, "Changing needs for Ph.D.'s," *IEEE Spectrum,* p. 79, Jan. 1972.

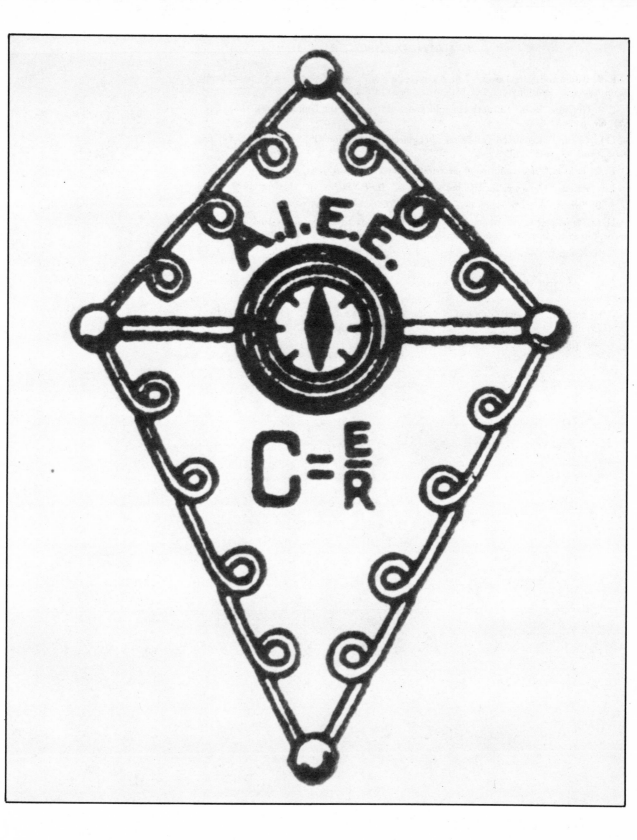

# 12

# AIEE + IRE = IEEE

As the title of this chapter indicates, in January 1963, the IRE merged with the AIEE and the two societies became the Institute of Electrical and Electronics Engineers. This chapter tells the story of how the merger came to pass and how the IEEE has fared during its first two decades. To set the tone of the story, it is well to record that the merger was welcomed by a large majority of the members of the Institutes. The vote in favor was the same in both groups: 87 percent. The voter response was large by society standards, as more than 60 percent of the eligible voters cast ballots.

There was, of course, opposition to the merger in both camps. In particular, the members of the AIEE Power Division feared that their interests would be overshadowed when the radio engineers of the IRE joined hands with their wire communication brethren in the AIEE. But as the years have passed, such fears have proved groundless.

On many counts — membership growth, the vitality of the IEEE Societies (notably the Society serving the interests of electric power), and the respect accorded the IEEE by other organizations and by industries and governments — the profession has profited from the merger. The status of the IEEE has steadily risen as electrical engineering and electronics have shown their abilities to embrace new technologies and to apply them in support of other areas, as with computers and electronic instrumentation.

If size is a significant measure, it is worth noting that the IEEE is now larger than any of the professional societies in the United States save one, the American Bar Association. In 1983, the IEEE membership approached a quarter of a million.

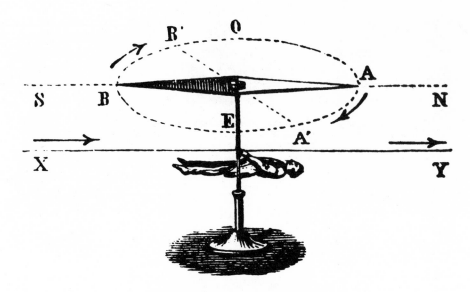

# WAY BACK IN 1912

When the IRE was formed in 1912, there was an opportunity for it to be welcomed as an equal partner by the AIEE. But the new field of radio could hardly be equated to the then-established activities in electric power and wire communications, so the AIEE failed to offer such a partnership. After consultation with AIEE Secretary Ralph Pope, the IRE founders decided to go it alone rather than accept junior status, and that chance for a unified society was lost. Historians have marked this a major mistake on the AIEE's part in view of the subsequent development of radio engineering and electronics.

There were, it is true, a few early events pointing toward greater cooperation between the AIEE and the IRE. In 1922, Dr. A. N. Goldsmith (IRE President, 1928) and A. E. Kennelly (AIEE President, 1899, and IRE President, 1916) discussed the possibility of merger, but without concrete result. A joint national convention was proposed for San Francisco in 1939, but there were local objections and only a few joint sessions were held. Joint sessions were held at the AIEE winter meeting and IRE convention in New York in the late 1940's. Standards were successfully handled jointly where the fields interlocked, and in later years radio standards were established by the IRE. But the climate was largely, "You do your job, we'll do ours."

We, the authors of this book, count it fortunate that the AIEE and the IRE remained separate for so many decades. This interval provided time for the development of two operational strategies, two radically different forms of technical organization, and two types of support for students. Above all it gave the IRE a chance to pursue a different aim for a technical society: the development of an engineering field based on scientific principles. The distinction between industry applications and science has by no means been sharply defined, but we believe that the record shows the major emphases of the Institutes were divided by that semantic barrier.

The AIEE's direction was dictated by the philosophy, "Science deals with physical laws, and engineering deals with the application of that knowledge to human needs." The IRE made no such distinction and welcomed physicists, metallurgists, chemists, and others from related scientific disciplines, provided only that they were working in fields allied to electronics. The IRE was thus ready at any time to move into new fields, and particularly ready for solid-state physics and the metallurgy, chemistry, and optics that brought forth the transistor and the integrated circuit chip.

In what follows, we have selected from the records of the AIEE and the IRE the principal data that reveal their strengths and weaknesses as they progressed toward the merger. This has been a difficult task; comparisons are all too often invidious. Doubtless many active members of either or both Institutes will take exception, if not to the facts themselves, then to our organization and interpretation of them. To members thus offended, we offer our apologies, but we have attempted to fairly present the issues faced, the measures taken, and the environments in which the two Institutes found themselves as they prepared to join hands.

# COMMON OBJECTIVES— DIFFERENT ROUTES

The two societies shared the common objective of serving qualified members in increasing numbers as the profession grew. The members' dues and their potential buying power as readers of advertisements provided the financial base of operations. The attractions of membership in a technical society include 1) receiving general and specialized publications that help one to keep abreast of new technology as well as news

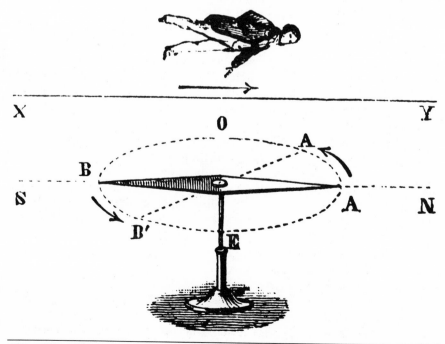

of the field; 2) attending technical conferences at which papers are presented and personal contacts established; 3) a geographical organization of local sections and chapters to provide for technical interchange without major travel costs; and 4) the privilege of wearing the badge and the recognition of excellence by one's peers.

To achieve these ends requires a headquarters organization, a publications staff, a regional organization to support the local sections, and committees to guide the many staff functions.

From their beginnings, the AIEE and the IRE achieved their objectives by traveling different routes, the principal difference being in organizational philosophy. The AIEE favored a centralized structure with power concentrated in the Board of Directors and its committees. The IRE, in part because it was based on rapidly changing technology and hence could not be controlled from the top, favored a decentralized structure. As time went on, the pace of change in electrical engineering became so rapid — and its proliferations so pervasive — that the decentralized approach was found necessary by both organizations before they merged.

# TWO ROUTES IN
# PUBLICATIONS

The different routes were particularly evident in the arrangements for handling publications and the technical conferences. The AIEE Technical Committee structure had its beginnings with two committees under President Scott in 1903. By 1945 there were 21 committees; the list had expanded to 50 by 1962. The technical committees had sole charge of accepting or rejecting papers for presentation at regional or technical conferences, and a paper had to be presented before it was published in the AIEE TRANSACTIONS. Discussion of the paper was published with the paper.

Later, the technical committees were grouped into six divisions, which included those covering electrical applications in industry, electric power generation and distribution, communications, and science and electronics. AIEE headquarters staff prepared three journals from papers supplied by the technical committees and also published ELECTRICAL ENGINEERING for the general membership.

The IRE technical structure largely centered on standards, but there were also committees for executive, financial, awards, membership, and other functions. In its early years, the IRE had only one publication, the PROCEEDINGS OF THE IRE; under Editor Alfred N. Goldsmith, this soon became a distinguished archival journal. Prior presentation of a paper was encouraged, but not required for publication. The AIEE TRANSACTIONS and the IRE PROCEEDINGS were refereed by expert editorial boards, and the quality of both was, in general, excellent. However, the centralized versus decentralized control of publication was clearly evident in the speed with which papers reached print. The IRE's policy of rapid publication was based on its science roots, since an early publication date was often sought to establish priority for new principles and discoveries.

The PROCEEDINGS did publish reviews and tutorial papers, and its special

Cover from the October 1926 issue of the
PROCEEDINGS OF THE IRE.

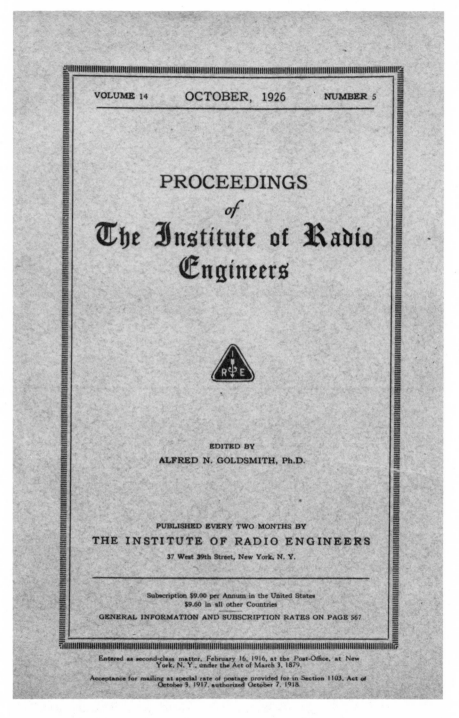

VOLUME 14     OCTOBER, 1926     NUMBER 5

# PROCEEDINGS
## *of*
# The Institute of Radio Engineers

EDITED BY
ALFRED N. GOLDSMITH, Ph.D.

PUBLISHED EVERY TWO MONTHS BY
### THE INSTITUTE OF RADIO ENGINEERS
37 West 39th Street, New York, N. Y.

Subscription $9.00 per Annum in the United States
$9.60 in all other Countries
GENERAL INFORMATION AND SUBSCRIPTION RATES ON PAGE 567

Entered as second-class matter, February 16, 1916, at the Post-Office, at New
York, N. Y., under the Act of March 3, 1879.

Acceptance for mailing at special rate of postage provided for in Section 1103, Act of
October 3, 1917, authorized October 7, 1918.

issues on current topics were of general interest. However, its usual papers were
too specialized and too mathematical to be understood except by a particular
audience. One joke explained that since the average reader could not understand
the technical papers, he had much time to peruse the advertisements, which

were indeed educational! Regardless of cause, the PROCEEDINGS advertising was a major source of IRE revenue after World War II. The IRE had in its PROCEEDINGS the best of two journalistic worlds: a respected journal and a steady source of income.

## THE IRE "SHOW"

Another difference, which had strong financial overtones, was the conduct in postwar years of the IRE Spring Convention, a grand "show." The AIEE did not undertake such an activity until just before the merger. The IRE event, held annually in March in New York City, displayed manufacturers' wares in a show open to members and to outsiders for a fee. These events were so well attended and so financially successful that IRE became a target of the U.S. Internal Revenue Service. The IRS claimed that the IRE was engaging in a form of profit-making not permitted by its tax-exempt status under paragraph 501(c)3 of the tax code. IRE countered that the purpose, function, and effect of its exhibitions were entirely educational.

The contest between the IRS and the IRE was not resolved until several years after the merger, when the IRS withdrew its complaint and rewrote the rules governing such activities by professional societies. One of the principal impediments to the merger was the tax liability that IRE had by then accrued, of the order of two million dollars. It was only after an inspection of IRE's books by the AIEE members of the Merger Committee, when IRE's reserves were found sufficient to cover the liability and more, that the merger discussion could proceed.

## PHILOSOPHIES FOR THE MEMBERSHIP GRADES

Qualifications for the various membership levels differed; the AIEE was more conservative, the IRE more liberal. From another viewpoint the performance requirements were tighter at the entrance level for the AIEE, but stricter at the top level for the IRE. The AIEE had Associate, Member, and Fellow grades; the IRE paralleled these with Associate, Member, Senior Member, and Fellow. In most respects, the AIEE Associate and Member grades bracketed the IRE Member and Senior Member levels.

Educational requirements for the AIEE were fixed at the level of colleges having curricula accredited by the Engineering Council for Professional Development (ECPD), whereas the IRE compiled its own list of "schools of recognized standing," including ECPD curricula and also others strong in the physical sciences, but not necessarily including engineering.

The AIEE Associate grade required ability to perform at the level of a graduate of an ECPD school, whereas the IRE Associate was open to anyone professing an interest in radio engineering, electronics, or any of the sister sciences. IRE Associates could not vote or hold office.

The AIEE grade of Fellow was available on application by Members with records of distinguished achievement. Among the AIEE Fellows were many of the highest distinction, everyone identifiable as an engineer. Unlike the AIEE, the IRE Fellow grade was conferred on those of outstanding achievement by action of the Board of Directors, following recommendation from the Fellow Committee. IRE conferred the Fellow grade on physicists, and in 1955 an award was made to Dr. William Shockley, who a year later shared the Nobel Prize for invention of the transistor. It went even further afield in 1962, giving the Fellow award to Dr. Grace Hopper, who was an outstanding figure in the development of computer software, having conceived the compiler method of organizing computer instructions. She was neither an engineer nor a physical scientist.

# DIFFERING VIEWPOINTS ON INTERNATIONALITY

Still another difference between the AIEE and IRE viewpoints deserves attention. From its founding, the AIEE thought of itself as primarily an American institution serving American engineers, and it placed that adjective in its name. The IRE, on the other hand, had from its beginnings placed emphasis on the transnationality of science and electrical technology. Therefore, no "A" appeared in the IRE's initials, and its constitution explicitly stated that its aims and geographical extent were worldwide. This was another area that had to be reconciled when the merger was being forged.

# THE 1947–1962 MEMBERSHIP RACE

The AIEE had started earlier and had served its constituents well. The IRE prospered when the electronics age arrived prior to World War II, but few foresaw the meteoric climb of the electronics industries after the war. Fortunately, in 1945, the IRE Board of Directors listened to urgings from Dr. F. E. Terman and Dr. W. L. Everitt and agreed to employ George W. Bailey as Executive Secretary. He came from the shoe industry by way of a wartime position as assistant to Vannevar Bush in the Office of Scientific Research in Washington. Bailey's job was to organize a headquarters staff. In the following years, Dr. W. R. G. Baker of G.E. (IRE President, 1947) and Bailey collaborated in guiding and guarding the finances of the growing IRE.

The next step was to bring the IRE into the view of the profession — out of rented space in the McGraw-Hill building, and into a converted mansion at 1 East 79th Street, New York, purchased with funds contributed by members and by the radio industry.

Thus, by 1947, the IRE was poised to ride the wave of the postwar growth of electronics. It would have been hard for the IRE to make a mistake that would have diverted it from that opportunity to grow. In fact, the IRE made few

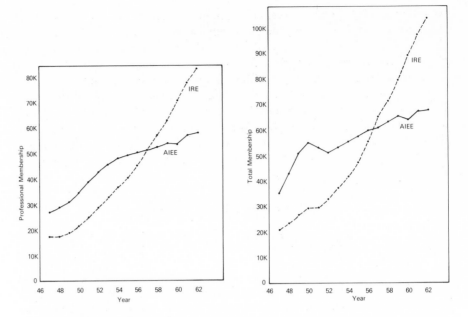

*Fig. 12.1 (left): Professional membership in the IRE and AIEE from 1947 to 1962.*
*Fig. 12.2 (right): Total membership in the IRE and AIEE from 1947 to 1962.*

mistakes and it did take actions that permitted its professional membership, to exceed that of the AIEE by 1957 (Fig. 12.1). In 1947, their total memberships numbered 35 491 in the AIEE and 21 037 in the IRE. Fig. 12.2 shows the trend in total membership during the 1947–1962 period.

The early comfortable margin of the AIEE over the IRE disappeared in part because the IRE was riding the electronics wave, but also because the IRE started its Professional Technical Groups in 1948 and the IRE STUDENT QUARTERLY in 1954. The latter was to have a strong impact on the recruitment of Student Members. The IRE Groups were to become the base of IEEE's present strong position in every phase of electrical engineering, electronics, and the allied arts and sciences.

# THE IRE GROUP PLAN

In 1948, the IRE Board heard a proposal by Raymond Heising of the Bell Laboratories that the Institute organize groups of members in particular disciplines. These groups would not only meet the needs of members in the established fields of electronics, but would attract as new members those entering new fields as those fields came into being. The IRE Board approved the plan and the IRE Professional Group System was established. Dr. W. L. Everitt was the first chairman of the Professional Groups Committee, followed by Dr. W. R. G. Baker, who was then the IRE Treasurer. The first IRE Professional Group was the Group on Audio Engineering, now the IEEE Society on Acoustics, Speech, and Signal Processing.

The IRE Group plan (adopted in all its essentials by the IEEE at the time of the merger) allowed any group of 100 or so members to petition for recognition as a subsociety within the IRE to cover a stated specialty area in the general field of interest to the IRE, including the related arts and sciences. The Group's Chairman and its Administrative Committee were elected by the Group members, and a subsidy for the Group's operations was provided in its formative years from headquarters funds. Each Group was encouraged to publish a journal, with its own Editor, and collectively these became known as the IRE TRANSACTIONS.

The Groups were permitted to organize their own conferences, with exhibits if desired. Profits from conferences and exhibits remained in the Group treasury. Some of the IRE Groups cooperated with AIEE Technical Committees in joint conferences prior to the merger. The Group system allowed expansion of services to members and Group fees provided additional funds to carry on those services.

Today, the IRE Groups and many of the AIEE Technical Committees are found among the IEEE Societies. The change in name from Group to Society came gradually; it was deemed to confer higher status.

In retrospect, the IRE Group system attracted new members from the enlarging pool of professionals entering special fields in electronics. Under IEEE auspices, the trend continues on a grander scale and today we find the IEEE Computer Society with over 70 000 members, the IEEE Power Engineering Society with some 20 000, and a total of 32 Societies (see Appendix IV).

# THE STUDENTS ARE
# HEARD FROM

On campus, the students were seeking recognition as members of a profession as early as the beginning years of the new century. The AIEE, under the urging of President Scott, established Branches as campus organizations in 1903.

In 1904, two electrical engineering students at the University of Illinois conceived of a national electrical engineering scholastic honor society. Late that year, the group had grown to ten, and Eta Kappa Nu was formed, using the Wheatstone bridge as its symbol. Today this arm of the profession is international in scope, with more than 160 Chapters worldwide, giving recognition to the best electrical students on each campus; the society also annually recognizes a young electrical engineer for achievements in his or her early professional years.

The IRE was late in providing for organizations of prospective members, not having Student Members until 1941. Only in 1947 were the first IRE Student Branches authorized, by which time the AIEE had 126 such groups. AIEE Branches were located only on ECPD-approved campuses; the IRE also allowed Branch location in schools having physics departments with strong electronics options.

The academic people in both Institutes insisted that electrical engineering and electronics were one field, based on one frequency spectrum and one set of electromagnetic laws. However, many schools established separate AIEE and IRE Branches, and their students were confronted with the necessity of joining two Branches if they wanted to partake of the full range of campus activities. This required the payment of two membership dues. The faculty counselors petitioned for the establishment of joint activities, with only one dues payment to the Institute of the student's choice. In 1950, the Boards of the two Institutes acceded to this request. By 1962, there were 130 Joint Student Branches, miniature versions of the monolithic merger still to come.

# THE MEMBERSHIP RACE
# AT THE GRASS ROOTS

The years 1947–1962 may be considered the formative years of the AIEE–IRE merger, starting with the authorization of IRE Student Branches in 1947 and ending with the merger itself. As shown in Fig. 12.3, student membership in both societies had peaked in the late 1940's, after which a government prediction of a severe oversupply of engineers had a depressing effect. IRE's recovery began in 1953, and in 1955 it overcame the AIEE lead. In part, this occurred because many young people were attracted to the growing fields of electronics, but another powerful force was at work. This was the recognition by the IRE of the importance of student enrollment to future membership, and the establishment in 1954 of the IRE STUDENT QUARTERLY.

Credit for the strong impact of this new publication clearly goes to T. A. "Ted" Hunter, a member of the IRE Board of Directors from Iowa City. He persuaded the Board to establish a journal for the students; his motto was "Ask the students!" and he used student advisors and reviewers. The STUDENT QUARTERLY was an immediate success and eventually achieved a circulation second only to that of the IRE PROCEEDINGS. Many IRE Members, long graduated, found it of such interest that they paid an extra fee to subscribe to it.

Hunter further persuaded the IRE Board to set up a permanent staff position

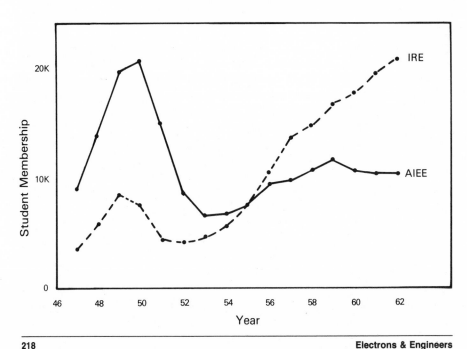

*Fig. 12.3: Student membership in the IRE and AIEE from 1947 to 1962.*

to support student programs. His attention to the student member was emotional, but also practical, because he saw that the future of any professional society rests on those entering it from the schools.

In 1959 the AIEE also started a student publication, the EE DIGEST, edited in New York by the staff, but it did not achieve the strong student support that went to the IRE QUARTERLY.

# THE AIEE REACTION

In 1946, the AIEE Board had written what had become known as the "Asheville Resolution" as a prescription for a postwar reorganization of the Institute. It reaffirmed the reliance on the centralized Technical Committee structure, although it allowed more flexibility in forming additional committees as the need arose. This led to an expansion of the number of Technical Committees from 21 to an eventual 50. Their coverage of technology was wide, but their contact with the interests of individual members was remote. Nowhere in evidence was the decentralized structure of the IRE Group System.

The AIEE Board set up a Committee on Planning and Coordination under the chairmanship of L. F. Hickernell (AIEE President, 1958–1959). Hickernell pointed to the slackening of AIEE's growth pattern that began in 1954. This was to have a particularly adverse effect on AIEE's plans, since half of its income came from members' dues. The factual record is shown in Figs. 12.1, 12.2, and 12.3. These trends were the cause for increasing concern by the AIEE Board of Directors.

In December 1957, AIEE President Walter Barrett of the New Jersey Bell Telephone Company, reacting to this concern, appointed a special task force,

---

*Left: Lloyd V. Berkner.*
*Right: Clarence H. Linder.*

with Prof. W. A. Lewis of the Illinois Institute of Technology as its chairman, to establish a "dynamic plan to accomplish objectives" pertinent to the 1957 situation. The Lewis Task Force reported in 1959. Its principal findings were that the AIEE had failed to enter new fields as they developed, that it did not appeal sufficiently to the students concerned with these new fields, and that the Board and its committee structure did not adequately cover the whole field in which electrical engineering was then engaged. It recommended that the AIEE establish a national technical group system, each group to be devoted to the activities of a parent Technical Committee. The similarity to the IRE Group System was evident and intentional, as Bill Lewis was also an IRE member and thoroughly familiar with its operations.

A plan to activate AIEE Institute Technical Groups (ITG's) was put before the membership in January 1961. By that time, the plans for the merger with the IRE were well underway, and the AIEE plan was never implemented. But when the Merger Committee came to the question of adoption or rejection of the IRE Group System, the groundwork for acceptance had been laid by the Lewis Task Force.

# THE MERGER LOOMS

Preceding the merger events were several talks between AIEE Presidents and their opposites in the IRE. One of these occurred in 1955 at the Engineers Club in New York between IRE President John D. Ryder and AIEE President Morris D. Hooven. From their luncheon conversation came a plan to permit transfer from membership in one Institute to the equivalent grade in the other without fee and without supporting documentation. The plan was later approved by both Boards.

Another luncheon at the Engineers Club took place in 1958, between

*Left: B. Richard Teare.*
*Right: Patrick E. Haggerty.*

Presidents L. F. Hickernell of the AIEE and D. G. Fink of the IRE. From their talk came further plans for collaboration, two points being joint Standards activity and joint Section meetings.

The record (and the recollections of those present) of several AIEE Board meetings show that the Board discussed the possibility of merging, but there is no record of an approach to the IRE. However, in January 1961, IRE President Lloyd Berkner arranged for Ronald McFarlan (IRE President, 1960) to appear before the AIEE Board, where he described IRE's organization and philosophy. Two months later, Clarence Linder, then AIEE President and Vice President for Engineering of G.E., appeared before the IRE Board, specifically to discuss AIEE's situation and the advisability of joint action by the two societies. Those present recall that he candidly stated that the membership trends of AIEE were not favorable to its future well-being. There was agreement that the professional interests of the members called for closer cooperation.

In April 1961, Berkner asked Patrick Haggerty (IRE President, 1962), Chairman of Texas Instruments, Inc., to confer with AIEE representatives. Haggerty reported progress to the IRE Executive Committee in May, including an agreement with Dr. B. R. "Dick" Teare (AIEE President, 1962–1963) that an eight-man joint AIEE–IRE Merger Committee would be set up. In the meantime and throughout the negotiations, letters from the Presidents were keeping the Sections informed. The joint Committee met in September and established a timetable of events: a constitution and bylaws to be ready for approval of both Boards in February 1962, with a membership vote by May of that year: if the vote succeeded, the merger was to occur on January 1, 1963. Some have called this haste, but overlooked are the differing official years of the Institutes: January to December for the IRE, July to June for the AIEE. Thus, a July 1962 deadline was imposed on merger negotiations by the constitutional need for the IRE to nominate a slate of 1963 officers for vote in September, if the merger was not to be completed.

# HOW IT WAS DONE

In late 1961, the Merger Committee was enlarged to 14 members (Appendix V), and the Committee attacked the problem of presenting to the two memberships by summer 1962, a specific recommendation, including a new constitution and bylaws. Many options had to be thrashed out: Was the new Institute to be transnational? Was it to employ the IRE Group System or the AIEE Technical Committee structure? How would the geographical Regions and Districts be adjusted? Would there be an Executive Committee for month-to-month decisions?

The Merger Committee met frequently and made decisions they hoped were appropriate for the future Institute. A constitution and bylaws were written and a ballot prepared for membership vote. After much discussion, the two Boards approved an amended version, and this went out for vote. The results, as reported earlier in this chapter, left no doubts for the future.

What, indeed, did the members approve? First, of course, they approved the merger itself. But they also approved a constitution and a set of bylaws. Here the record is clear and — to the authors of this book — misleading. It takes only a reading of the IEEE, AIEE, and IRE documents to learn that the IRE pattern had prevailed. Two explanations come to mind, neither of which grasps the inner significance of the event. One is that the IRE representatives were unwilling to consider the merger if IRE's traditions and modi operandi were changed. The other is that AIEE approached the merger from a weakened position, financially and in membership trends. This view further indicates that the AIEE had already reached the understanding within its own councils that the centralized Technical Committee structure, with its firm control of AIEE's technical affairs, was no longer able to match the pace of change in electrical engineering.

Such arguments were heard and there is truth in them, but they fail to reveal the true reason for the merger: both societies were incomplete. Neither Institute could claim on its own to cover the professions of electrical engineering and electronics, to serve the industries involved, or to hold the allegiance of academia. Together they could cover the ground, but separately they could only engage in destructive competition.

It was this deeper view of the need for merger that impelled its two principal architects: Clarence Linder of the AIEE and Pat Haggerty of the IRE, ably seconded by Dick Teare and Lloyd Berkner, respectively. These men were statesmen, not men reacting to events. Linder worked as hard in convincing the IRE officers of the merger's merits as he did with his fellow officers in AIEE. Haggerty had the facility for presenting ideas clearly and forcefully, and his conviction spread through the IRE camp and spilled over into the AIEE.

# THE NAME AND BADGE

Later in the discussion by the 14-man Merger Committee came the choice of the new Institute's name and the design of a badge. It was agreed that the proposed Institute would be transnational in character, serving electrical engineers around the world, and the "A" in AIEE was removed. That left "Institute of Electrical Engineers," but the initials IEE

had already been preempted by the British Institution of Electrical Engineers. Some of the IRE contingent suggested that "electronics" should be recognized in the name, despite the general awareness that electrical engineering included electronics — albeit as a most active and exciting subdivision.

There was thought that the IRE's devotion to scientific principles should be recognized and the "Institute of Electrical Science and Engineering" was suggested. IESE found some favor as descriptive of the field, but "Institute of Electrical and Electronic(s) Engineers" was the final choice.

The use of the "s" on electronics in the name was not decided until February 1963, following argument that an electronic engineer could only be a robot, operating by internal tubes or transistors. It was also noted that in English usage, fields of knowledge were designated by plurals as mathematics, physics, economics; thus, the profession practiced by the members was electronics. And so the inclusion of the "s" was decided.

In contrast, the design of the badge was settled in less than 15 minutes. It was perceived that the AIEE and IRE badges had features that could be combined. The four-sided "kite" outline of the AIEE badge was retained, with minor adjustment. The central feature of the IRE badge, the straight and curved arrows symbolizing the right-hand rule of electromagnetism, was recognized as basic in electrical theory from megawatt power generators to radio waves. One change was to have the central arrow point upward (*ad astra*) rather than down as in the IRE symbol.

All this was quickly absorbed by one member of the committee, Dr. Bernard M. "Barney" Oliver (IEEE President, 1965), Vice President of the Hewlett-Packard Company and handy with a drawing pencil. He sketched out the badge as it had been described, to everyone's satisfaction.

One notable difference between the proposed badge and its predecessors was the absence of lettering. Nelson Hibshman urged this change, and it was taken up in the Merger Committee by Jack Ryder, who pointed out that a badge without letters could be read in any language. Oliver added that letters would clutter the design. Beneath this light-hearted banter was an implied hope that, in time, the outline of the badge would become so familiar in engineering and scientific professions that further identification would not be needed.

*Left: The first badge of the AIEE. The shape is Franklin's kite, the border is a Wheatstone bridge, the center is a galvanometer, and Ohm's law is described.*
*Center: This AIEE badge design supplanted the original in 1897. The kite has been modified; the linked circles represent the interdependence of electricity and magnetism.*
*Right: The IRE adopted a simple badge design from the outset, with the right-hand rule at the center.*
*Above: The badge of the IEEE needs no identifying letters. The arrows of the right-hand rule say it all.*

# THE FIRST
# GENERAL MANAGER

The specifications for a man to head the IEEE's headquarters staff were complex. It was realized that the new Institute would be primarily a publishing house, as the two merging Institutes were publishing 39 journals. The combined AIEE and IRE budgets exceeded five million dollars, thus requiring financial ability. It was too much to ask that a single individual be eminent in both power and electronics, but he should be eminent in one and preferably be an AIEE or IRE Fellow. Pat Haggerty insisted that the new man have management experience and serve as General Manager, not Executive Secretary as George Bailey and Nelson Hibshman were designated in the IRE and the AIEE.

Haggerty led the search and contacted Don Fink in June 1962. He was a Fellow of both Societies and a past President of the IRE, had from 1952 been Director of Research for Philco Corporation, and in 1962 became Director of Philco-Ford Scientific Laboratories. Also, he had been Vice Chairman of the National Television System Committee that set up U.S. color television standards. During his interview with the Merger Committee, he was welcomed by most of the AIEE group as "one of our boys" after he was revealed as an AIEE Fellow, although the power representatives were doubtful. When Fink later asked for permission to serve as Editor-in-Chief of McGraw-Hill's *Standard Handbook for Electrical Engineers*, the bible of the power profession, he was fully endorsed by the power contingent. He was appointed, effective with the merger, January 1963.

Since Fink's retirement in 1974, the General Manager post has been filled by Herbert A. "Judd" Schulke, retired Major General of the U.S. Army; Dr. Richard M. Emberson; and the 1984 incumbent Eric Herz.

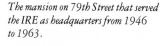

*The mansion on 79th Street that served the IRE as headquarters from 1946 to 1963.*

# THE IEEE IS BORN

On the legal side, it was agreed without rancor that the AIEE would be the continuing organization, that the IRE would merge with it and disappear as a corporation, and that the AIEE would change its name to IEEE. Since both the AIEE and the IRE had been New York corporations, this was a New York legal process, and the renamed corporation was officially authorized to do business as of the first week of January 1963.

Dr. Ernst Weber, then President of the Polytechnic Institute of Brooklyn (IRE President, 1959), was so well respected in AIEE affairs that there was no question of partisanship when the Merger Committee chose him for the first President of the IEEE. The Merger Committee also proposed the first Board of Directors, choosing equally from among the top talent in former officers and directors of the two predecessor societies.

What lay before the Board and the staff was a mammoth job of compromise to bring into being a single society from two different preceding cultures. Initially, several of the AIEE Divisions chose to remain outside the new IEEE Group structure. One was the AIEE Power Division; it was not until 1965 that the IEEE Group for Power Equipment and Systems was formed and took over the Power Division's responsibilities.

Another matter requiring attention was the disposition of headquarters facilities. The AIEE had, in 1961, moved into the newly opened United Engineering Center on 47th Street, near the United Nations building. The proximity of these quarters to the other engineering societies had indicated that the IEEE Headquarters should be at 47th Street. But the IRE had a large investment in its buildings at 79th Street and Fifth Avenue, a collection of former mansions that were beautiful in architecture and furnishing and to which the former IRE staff had developed a strong attachment.

In 1964, when Clarence Linder was IEEE President, a buyer was found for the IRE property at what was considered a fair price. The sale was made and the staff

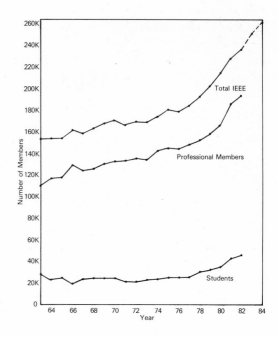

*Fig. 12.4: Growth of the IEEE from its birth in 1963 to the present.*

was transferred to 47th Street. Time and air-conditioned comfort took care of the nostalgia, and the staff quickly became as happy and cohesive a group as can be found in professional society headquarters. Aiding in this were Fink's recommendations to the Merger Committee that staff reduction occur only by attrition, and that wherever employment practices differed, the more generous be adopted by the IEEE.

In 1983, there were a number of active staff members whose careers started in the AIEE or the IRE before the merger. The record for longevity is held by Thomas W. "Tom" Bartlett, the IEEE Controller, who started with the AIEE in 1946 as a clerk in the shipping room. Second in longevity came from the IRE in the person of Elwood K. "Woody" Gannett, the son of an IRE Fellow, who also joined the staff in 1946, just out of engineering school and the U.S. Navy. He is now Deputy General Manager, having previously served as Staff Director of Publishing Services. Another legendary figure, probably known to more IEEE members than any other, was Emily Sirjane, now deceased, on whom the Board conferred the title of "Honorary Fellow."

There was some education required at the Board level, one lesson revolving around the question of whether the General Manager was to be permitted to handle administrative matters without overrule by the Executive Committee. That point was resolved in favor of management authority, but both sides learned from the argument.

The steady growth of IEEE membership (Fig. 12.4) and the consequent requirements for increased staff put a strain on the space available to the IEEE at the Engineering Center. After expansion to two additional floors, Fink was authorized in 1971 to determine the feasibility of building an IEEE Service Center in Piscataway, NJ. Although there was unhappiness among staff faced with moving families, and the project was known in some quarters as "Fink's Folly," the center was constructed and occupied in 1973. Today it is highly successful and an excellent investment, the land alone having increased in value one hundredfold.

# THE BIRTH
# AND GROWTH OF
# *IEEE SPECTRUM*

At the time of the merger, there were two publications sent to members as a basic perquisite of membership: AIEE's ELECTRICAL ENGINEERING and the IRE PROCEEDINGS.

ELECTRICAL ENGINEERING had been edited at AIEE headquarters by staff. From its founding in 1914, the IRE PROCEEDINGS had been directed by a volunteer Editor who was a director and officer of the Institute. For 39 years, to 1953, this volunteer was Dr. A. N. Goldsmith, one of the IRE's founders, and credit for the high standing of the PROCEEDINGS as an archival journal goes to him. Thereafter, the post of IRE Editor was held for short terms by John R. Pierce, D. G. Fink, John D. Ryder, Ferdinand Hamburger, Jr., and Thomas F. Jones, Jr.

Ryder, as a member of the Merger Committee, was asked to explain the functioning of the post of IRE Editor to the AIEE members of that committee, after which they accepted the post for the organization chart and appointed him IEEE's first Editor. It fell to him and his Editorial Committee to decide the future of the two previous general publications and to formulate policies for a core journal for all IEEE members.

*The first issue of* SPECTRUM, *January 1964.*

Financially, the PROCEEDINGS had been a source of revenue for the IRE, but ELECTRICAL ENGINEERING had not attracted enough advertising to support itself. A consultant, Ralph Flynn (former publisher of *Electrical World*) was hired. He advised against continuing ELECTRICAL ENGINEERING, but felt that the PROCEEDINGS should be retained as an archival journal available to members by subscription. This left a vacuum to be filled by a new publication for all IEEE Members.

At a meeting with Flynn in Fink's office, Ryder described his view of the journal: it should be a warm magazine that carried IEEE news and words from its leaders, with technical content aimed just above the median of the IEEE membership, so as to pull the Members along without swamping them. That prescription was adopted by the IEEE Board, and IEEE SPECTRUM was born in January 1964.

The name SPECTRUM had been advanced for the student publication by W. Reed Crone, Staff Editor of the STUDENT QUARTERLY, but was yielded gracefully after it was realized how appropriate the word was for the field of interest of the IEEE.

Editor Ryder and Woody Gannett of the staff made Society publication history by hiring a professional writer for special assignments and an art editor to produce color covers and layouts of professional quality. Ryder turned over his post as IEEE Editor in 1965 to F. Karl Willenbrock. In 1966, the volunteer post of SPECTRUM Editor was established, and this was held by C. Chapin Cutler, J. J. G. McCue, and David DeWitt.

By 1970, the Board of Directors agreed that the SPECTRUM editorship was becoming a management function, one too heavy for a volunteer to carry.

A selection of the 59 journals and magazines currently published by the IEEE.

General Manager Fink instituted a search for a full-time experienced Editor, and found such a person in Donald Christiansen, former Editor-in-Chief of the McGraw-Hill magazine *Electronics,* the position that Fink himself had held 20 years earlier.

Christiansen agreed to take the job, even though IEEE's salary could not then match what he was earning at McGraw-Hill. He wanted the challenge of editing a publication of great personal interest to him, one for which he saw important growth possibilities.

In retrospect, Fink is now of the opinion that if he had done nothing for IEEE but hire Christiansen, he would have been satisfied. Under Christiansen's leadership, IEEE SPECTRUM won a National Magazine Award in 1979 for its issue that covered the nuclear accident at Three Mile Island. In 1982, its issue on technology in war and peace showed how expert contributors can work with capable staff editors to produce a monument to technical journalism, winning the National Magazine Award once again, this time against such formidable contenders as *Time, Life, Business Week,* and *Scientific American.*

The IEEE STUDENT JOURNAL ceased publication in 1970, when the Board of Directors determined that SPECTRUM could fulfill the needs of the Student Members. But a student magazine was started up once again in 1982. With some poetic justice, the magazine was named IEEE POTENTIALS at the suggestion of SPECTRUM'S Editor — who had benefited, as noted earlier, when the original student magazine relinquished its planned title, SPECTRUM. IEEE POTENTIALS was designed and launched under the guidance of Christiansen.

*In 1983 the U.S. Postal Service honored the professionals of the IEEE with a set of stamps depicting four of the greatest: Steinmetz, Armstrong, Tesla, and Farnsworth.*

# THE IEEE GOES PROFESSIONAL

One of the first to remind IEEE Members that they were people as well as engineers and scientists was IRE founder Alfred Goldsmith, who wrote in the first issue of SPECTRUM: "It is obviously essential that an engineer and his family shall live an economically pleasing life free from harassment or distress arising from his daily needs and the development of his long-term security."

Not everyone agreed that such matters were the business of the IEEE. In 1964, the ad hoc Committee on Professional Relations recommended that "the distinction between activities appropriate to a technical society and the activities appropriate to organizations concerned with the legal, economic, and public relations aspects of the engineering profession should be preserved." However, by 1970, the atmosphere had changed, there was unemployment among IEEE members, and pensions were of concern, as patterns of government procurement were forcing some members to change employers so frequently that they never qualified for pension rights.

In 1971, the Board sent a questionnaire to members to ask if the IEEE should represent the profession in nontechnical matters. The response was positive, and the Board tackled the explosive issue of amending the IEEE Constitution to allow the Board to carry out the professional objectives.

Although some said this was to be a move toward unionization of the electrical profession, that action was, in fact, to be expressly barred. Actually, it was a move by the IEEE to join the large number of societies serving professions, notably medicine and law, that took an active part in any matter affecting "the good of the profession," including political involvement and lobbying before Congress and other governmental bodies.

One important change was to be in the IEEE tax status; previously, as a "501(c)3" society, it had enjoyed several privileges, including favorable postage rates and income tax credits for member dues. Societies that engaged in political or lobbying activities and matters affecting the self-interest of members were classed as "business leagues" under paragraph 501(c)6 of the tax code. Tax credits for dues and fees could be taken only as business expenses, and there were other restrictions, acceptable to doctors and lawyers, but in doubt for application to engineers and scientists.

The Board of Directors' debate was sharply divided, and finally Arthur P. Stern (IEEE President, 1975) asked General Manager Fink to compose a constitutional amendment that could be approved by the several factions.

This amendment retained the previous statement of IEEE directions as "advancement of the theory and practice of electrical engineering, electronics, radio, and the allied branches of engineering and the related arts and sciences," and the holding of meetings and publications pertaining thereto. The departure from the previous technical limits came in a second paragraph, in which the IEEE was also to be directed toward the advancement of the standing of the members of the professions it served, and spelled out permitted means to that end. In its conclusion, it specifically excluded from IEEE purview and action matters related to "collective bargaining, as salaries, wage, benefits and working conditions, customarily dealt with by labor unions."

To prepare the members for the vote in the fall of 1972, Fink wrote a paper describing the issues and the reasons for the Board's recommendation for favorable action. The amendment was approved by 87 percent of the membership. In its first implementing action, the IEEE applied for and received 501(c)6 tax status from the U.S. Internal Revenue Service on February 1, 1973. The IEEE thus embarked on new endeavors that exposed it to great risk or to the promise of great reward, depending on the wisdom to be shown by the Institute's leaders.

# PROFESSIONALISM
# AT WORK

The implementation of the new constitutional clause, in which the Institute involved itself with the welfare of its members, was essentially limited to the United States, where the large majority of IEEE members reside, although similar actions in other countries are permitted within their own legal confines. The U.S. Regions established a U.S. Activities Committee, later designated a Board (USAB), supported by an annual assessment paid only by U.S. resident Members. This Board now has a separate staff in Washington, DC, where other power centers governing many phases of science and engineering are located.

Two other results of the constitutional change are worthy of mention. The first was the formation of an IEEE Society having no specific technical charge, namely, the IEEE Society on the Social Implications of Technology. The second was a long campaign in the U.S. Congress to reform pension practices, namely, the lowering of the number of years required for vesting of pension rights. The IEEE and others also successfully pressed Congress for establishment of the Individual Retirement Account (IRA), under which individuals could set aside amounts of income without current income tax, for withdrawal later in retirement. Those members of Congress who were concerned acknowledge that the IEEE case for such reform was early and well presented.

# THE CENTENNIAL APPROACHES

After the constitutional change, the expansion of IEEE membership services has continued; its membership growth has accelerated as shown in Fig. 12.4.

One other change shows the arrival of a new maturity in the IEEE. Previously, there had been one instance of a successful petition candidate for President in each of the AIEE and the IRE, but in 1981, for the first time, the Board of Directors presented to the members a board-approved slate of two eminent candidates for each of the offices of President-Elect and Executive Vice President. As a result, the President serving the IEEE during its Centennial Year is Dr. Richard J. Gowen, a seasoned veteran of IEEE service from the academic ranks, who won a closely contested election against Dr. Donald D. King, a talented engineer then serving as head of a major research laboratory.

The earlier practice of offering only one Board-nominated candidate had been based on the maxim that governs many other societies: good men who can and will offer their services for these demanding positions are scarce. But now a point has been made without doubt: the IEEE has, and will continue to have through its Societies' training grounds, a sufficient supply of talented leaders equal to those who have served it so well in the past.

## For Further Reading

C. Dreher, "His colleagues remember the 'Doctor,'" *IEEE Spectrum*, p. 32, Aug. 1974.

J. Fagenbaum, "Patrick Haggerty: Engineer and visionary," *IEEE Spectrum*, p. 20, Dec. 1980.

D. G. Fink, "Blueprint for change," *IEEE Spectrum*, vol. 9, p. 38, June 1972.

E. Rubinstein, "IEEE and the founder societies," *IEEE Spectrum*, p. 76, May 1976.

J. D. Ryder, "Poles and zeros," *Proc. IEEE*, vol. 51, p. 295, Feb. 1963.

——, "The genesis of an editorial policy," *IEEE Spectrum*, vol. 1, p. 147, 1964.

*Special acknowledgment is given to the following:*

E. T. Layton, Jr., "Scientists and engineers: The evolution of the IRE," *Proc. IRE*, vol. 64, p. 1390, 1976.

A. M. McMahon, "Corporate technology: The social origins of the American Institute of Electrical Engineers," *Proc. IRE*, vol. 64, p. 1383, 1976.

# *Presidents of the AIEE*

| DATE OF OFFICE | NAME | AGE AT TAKING OFFICE | EDUCATION | BUSINESS ACTIVITY[1] |
|---|---|---|---|---|
| 1884–86 | Green, Norvin | 56 | Medicine | Mgr. |
| 1886–87 | Pope, Franklin L. | 46 | — | Attorney |
| 1887–88 | Martin, T. Commerford | 31 | Theology | Editor |
| 1888–89 | Weston, Edward | 38 | — | Mfr. |
| 1889–90 | Thomson, Elihu | 36 | High school | Mfr. |
| 1890–91 | Anthony, William A. | 55 | Yale '56 | Consult. |
| 1891–92 | Bell, Alexander G. | 44 | Edinburgh | Mfr. |
| 1892–93 | Sprague, Frank J. | 35 | USNA | Mfr. |
| 1893–95 | Houston, Edwin J. | 46 | High school | Mfr. |
| 1895–97 | Duncan, Louis | 34 | USNA | Prof. |
| 1897–98 | Crocker, Francis B. | 36 | Ph.D. Columbia '95 | Prof. |
| 1898–1900 | Kennelly, Arthur E. (Pres. IRE 1916) | 37 | Univ. Coll. London | Prof. |
| 1900–01 | Hering, Carl | 40 | Univ. Pennsylvania '80 | Consult. |
| 1901–02 | Steinmetz, Charles P. | 36 | Breslau | Consult. |
| 1902–03 | Scott, Charles F. | 38 | Ohio State '85 | Prof. |
| 1903–04 | Arnold, Bion J. | 42 | Nebraska '97 | Consult. |
| 1904–05 | Lieb, John W. | 44 | Stevens Tech '80 | Mgr. Util. |
| 1905–06 | Wheeler, Schuyler S. | 45 | Columbia '82 | Mfr. |
| 1906–07 | Sheldon, Samuel | 44 | Ph.D. Wurzburg '88 | Consult. |
| 1907–08 | Stott, Henry G. | 41 | Glasgow '85 | Mgr. Transp. |
| 1908–09 | Ferguson, Louis A. | 41 | M.I.T. '88 | Mgr. Util. |
| 1909–10 | Stilwell, Lewis B. | 46 | Lehigh '85 | Mgr. Util. |
| 1910–11 | Jackson, Dugald C. | 45 | Penn State '85 | Prof. |
| 1911–12 | Dunn, Gano | 41 | Columbia '99 | Consult. |
| 1912–13 | Mershon, Ralph D. | 44 | Ohio State '90 | Consult. |
| 1913–14 | Mailloux, C. O. | 53 | Columbia | Editor |
| 1914–15 | Lincoln, Paul M. | 44 | Ohio State '92 | Prof. |
| 1915–16 | Carty, John J. | 54 | High school | Mgr. Teleph. |
| 1916–17 | Buck, Harold W. | 53 | Columbia '95 | Consult. |
| 1917–18 | Rice, Edwin W., Jr. | 55 | High school | Mgr. |

| | | | | |
|---|---|---|---|---|
| 1918–19 | Adams, Comfort A. | 50 | Case '90 | Consult. |
| 1919–20 | Towsley, Calvert | 55 | Yale '88 | Mgr. |
| 1920–21 | Berresford, Arthur W. | 48 | Cornell '93 | Mgr. |
| 1921–22 | McClellan, William | 49 | Ph.D. Univ. Pennsylvania '03 | Consult. |
| 1922–23 | Jewett, Frank B. | 43 | Ph.D Chicago '02 | Mgr. Teleph. |
| 1923–24 | Ryan, Harris J. | 57 | Cornell '87 | Prof. |
| 1924–25 | Osgood, Farley | 50 | M.I.T. '97 | Consult. |
| 1925–26 | Pupin, Michael I. (Pres. IRE 1917) | 67 | Ph.D. Berlin '89 | Prof. |
| 1926–27 | Chesney, Cummings C. | 63 | Penn State '85 | Mgr. |
| 1927–28 | Gherardi, Bancroft | 54 | Cornell '94 | Mgr. Teleph. |
| 1928–29 | Schuchardt, Rudolph F. | 43 | Wisconsin '97 | Mgr. Util. |
| 1929–30 | Smith, Harold B. | 38 | Cornell '91 | Prof. |
| 1930–31 | Lee, William S. | 58 | Citadel '94 | Mgr. Util. |
| 1931–32 | Skinner, Charles E. | 66 | Ohio State '90 | Mgr. |
| 1932–33 | Charlesworth, Harry P. | 50 | M.I.T. '05 | Mgr. Teleph. |
| 1933–34 | Whitehead, John B. | 61 | Ph.D. Johns Hopkins '02 | Consult./Prof. |
| 1934–35 | Johnson, J. Allen | 52 | W.P.I. '05 | Mgr. Util. |
| 1935–36 | Meyer, Edward B. | 53 | Pratt '03 | Mgr. Util. |
| 1936–37 | MacCutcheon, Alexander M. | 55 | Columbia '08 | Mgr. |
| 1937–38 | Harrison, William H. | 45 | Pratt '15 | Mgr. Teleph. |
| 1938–39 | Parker, John C. | 59 | Univ. Michigan '04 | Mgr. Util. |
| 1939–40 | Farmer, F. Malcolm | 62 | Cornell '99 | Consult. |
| 1940–41 | Sorenson, Royal W. | 58 | Univ. Colorado '05 | Prof. |
| 1941–42 | Prince, David C. | 50 | M.S. Univ. Illinois '13 | Consult. |
| 1942–43 | Osborne, Harold S. | 55 | M.I.T. '08 | Mgr. Teleph. |
| 1943–44 | Fink, Nevin E. | 60 | Lehigh '05 | Mgr. Util. |
| 1944–45 | Powel, Charles A. | 60 | Bern '05 | Mfr. |
| 1945–46 | Wickenden, William E. | 63 | Denison '04 | Prof. |
| 1946–47 | Housley, J. Elmer | 53 | Univ. Tennessee '15 | Mgr. Util. |
| 1947–48 | Hull, Blake D. | 65 | Kansas '05 | Mgr. Teleph. |
| 1948–49 | Lee, Everett S. | 57 | Univ. Illinois '13 | Mgr. |
| 1949–50 | Fairman, James F. | 53 | M.S. Univ. Michigan '21 | Mgr. Util. |
| 1950–51 | LeClair, Titus G. | 51 | Univ. Idaho '18 | Mgr. Util. |
| 1951–52 | McMillen, Fred D. | 61 | Oregon State '12 | Prof. |
| 1952–53 | Quarkes, Donald A. | 58 | Yale '16 | Mgr. Teleph. |
| 1953–54 | Robertson, Elgin B. | 60 | Univ. Texas '15 | Mfr. |
| 1954–55 | Monteith, Alexander C. | 52 | Queen's '23 | Mgr. |
| 1955–56 | Hooven, Morris D. | 58 | Bucknell '20 | Mgr. Util. |
| 1956–57 | Coover, Mervin S. | 66 | R.P.I. '14 | Prof. |
| 1957–58 | Barrett, Walter J. | 58 | B.P.I. '20 | Mgr. Util. |
| 1958–59 | Hickernell, L. F. | 59 | M.I.T. '22 | Mfr. |
| 1959–60 | Foote, James H. | 68 | Michigan State '14 | Mgr. Util. |
| 1960–61 | Linder, Clarence H. | 57 | Univ. Texas '24 | Mgr. |
| 1961–62 | Chase, Warren H. | 63 | Harvard '24 | Mgr. Teleph. |
| 1962–63 | Teare, B. Richard, Jr. | 55 | Wisconsin '27 | Prof. |

[1] *An arbitrary classification: Mgr. covers a President, Vice President, Chief Engineer; Consultant covers cases of industrial employment where freedom of problem choice was given; Acad. indicates an Administrator at a university.*

233

# *Presidents of the IRE*

| DATE OF OFFICE | NAME | AGE AT TAKING OFFICE | EDUCATION | BUSINESS ACTIVITY |
|---|---|---|---|---|
| 1912 | Marriott, Robert H. | 33 | Ohio State '01 | Consult. |
| 1913 | Pickard, Greenleaf W. | 36 | M.I.T. | Consult. |
| 1914 | Austin, Louis W. | 47 | Ph.D. Strassburg | Res. |
| 1915 | Stone, John S. | 46 | Johns Hopkins | Consult. |
| 1916 | Kennelly, Arthur E. (Pres. AIEE 1898–99) | 55 | Univ. Coll. London | Prof. |
| 1917 | Pupin, Michael L. (Pres. AIEE 1925–26) | 59 | Ph.D. Berlin '89 | Prof. |
| 1918–19 | Pierce, George W. | 47 | Ph.D. Leipzig '00 | Prof. |
| 1920 | Hogan, John V. L. | 30 | Yale | Consult. |
| 1921 | Alexanderson, Ernst F. W. | 43 | Royal Tech I. Sweden | Consult. |
| 1922 | Cutting, Fulton | — | Sc.D. Harvard | Prof. |
| 1923 | Langmuir, Irving | 42 | Ph.D. Gottingen | Res. |
| 1924 | Morecroft, John H. | 43 | Syracuse '04 | Prof. |
| 1925 | Dellinger, John H. | 39 | Ph.D. Princeton '13 | Consult. |
| 1926 | McNicol, Donald | 51 | | Consult. |
| 1927 | Bown, Ralph | 36 | Ph.D. Cornell '17 | Res. |
| 1928 | Goldsmith, Alfred N. | 41 | Ph.D. Columbia '11 | Consult. |
| 1929 | Taylor, A. Hoyt | 50 | Ph.D. Gottingen '09 | Res. |
| 1930 | deForest, Lee | 59 | Ph.D. Yale '99 | Consult. |
| 1931 | Manson, Ray H. | 54 | Maine '98 | Mgr. |
| 1932 | Cady, Walter G. | 58 | Ph.D. Berlin '00 | Prof. |
| 1933 | Hull, Lewis M. | 34 | Ph.D. Harvard '19 | Res. |
| 1934 | Jansky, C. M., Jr. | 39 | M.S. Wisconsin '19 | Consult. |
| 1935 | Ballantine, Stuart | 38 | Harvard | Res. |
| 1936 | Hazeltine, Alan | 50 | Stevens '06 | Res. |
| 1937 | Beverage, Harold H. | 44 | Maine '15 | Res. |
| 1938 | Pratt, Haraden | 47 | Univ. California '14 | Consult. |
| 1939 | Heising, Raymond A. | 51 | M.S. Wisconsin '14 | Res. |
| 1940 | Horle, Laurence C. F. | 48 | Stevens '14 | Consult. |
| 1941 | Terman, Frederick E. | 41 | Sc.D. M.I.T. '22 | Prof. |

| 1942 | Van Dyck, Arthur F. | 51 | Yale '11 | Consult. |
|------|--------------------|----|----------|----------|
| 1943 | Wheeler, Lynde P. | 69 | Ph.D. Yale '02 | Consult. |
| 1944 | Turner, Hubert M. | 62 | Illinois '10 | Prof. |
| 1945 | Everitt, William L. | 45 | Ph.D. Ohio State '32 | Prof. |
| 1946 | LLewellyn, Frederick W. | 49 | Stevens '22 | Res. |
| 1947 | Baker, Walter R. G. | 47 | M.S. Union '18 | Mgr. |
| 1948 | Shackelford, Benjamin E. | 57 | Ph.D. Chicago '16 | Mgr. |
| 1949 | Bailey, Stuart L. | 54 | M.S. Minnesota '28 | Consult. |
| 1950 | Guy, Raymond F. | 51 | Pratt '21 | Consult. |
| 1951 | Coggeshall, Ivan S. | 55 | Worcester '17 | Res. |
| 1952 | Sinclair, Donald B. | 42 | Sc.D. M.I.T. '35 | Mfr. |
| 1953 | McRae, James W. | 43 | Ph.D. Cal Tech '37 | Mgr. |
| 1954 | Hewlett, William R. | 41 | M.S. M.I.T. '36 | Mfr. |
| 1955 | Ryder, John D. | 48 | Ph.D. Iowa State Univ. '44 | Prof. |
| 1956 | Loughren, Arthur V. | 54 | Columbia '25 | Consult. |
| 1957 | Henderson, John T. | 52 | Ph.D. London | Res. |
| 1958 | Fink, Donald G. | 47 | M.S. Columbia '42 | Consult. |
| 1959 | Weber, Ernst (Pres. IEEE 1963) | 58 | Ph.D. Vienna | Prof. |
| 1960 | McFarlan, Ronald L. | 55 | Ph.D. Chicago | Consult. |
| 1961 | Berkner, Lloyd V. | 56 | Minnesota '27 | Mgr. |
| 1962 | Haggerty, Patrick E. | 48 | Marquette '36 | Mfr. |

# *Presidents of the IEEE*

| DATE OF OFFICE | NAME | AGE AT TAKING OFFICE | EDUCATION | BUSINESS ACTIVITY |
|---|---|---|---|---|
| 1963 | Weber, Ernst (Pres. IRE 1959) | 62 | Ph.D. Vienna | Acad. |
| 1964 | Linder, Clarence | 61 | M.S. Univ. Texas '27 | Mgr. |
| 1965 | Oliver, Bernard | 49 | Ph.D. C.I.T. '40 | Res. Dir. |
| 1966 | Shepherd, William | 55 | Ph.D. Univ. Minnesota '37 | Acad. |
| 1967 | MacAdam, Walter | 54 | S.M. M.I.T. '37 | Teleph. |
| 1968 | Herwald, Seymour W. | 51 | Ph.D. Univ. Pittsburgh | Res. Dir. |
| 1969 | Willenbrock, Karl | 49 | Ph.D. Harvard | Acad. |
| 1970 | Granger, John V. N. | 52 | Ph.D. Harvard | Mfr. |
| 1971 | Mulligan, James | 51 | Ph.D. Columbia '48 | Acad. |
| 1972 | Tanner, Robert | 57 | M.S. Imperial Coll. | Res. Dir. |
| 1973 | Chestnut, Harold | 56 | M.S. M.I.T. '40 | Mgr. |
| 1974 | Guarrera, John | 52 | M.I.T. '43 | Mgr. |
| 1975 | Stern, Arthur | 50 | M.S. E.T.H. '48 | Mgr. |
| 1976 | Dillard, Joseph | 59 | M.S. M.I.T. '50 | Mgr. |
| 1977 | Saunders, Robert M. | 62 | M.S. '38 Minnesota | Prof. |
| 1978 | Getting, Ivan | 66 | Ph.D. Oxford '35 | Mfr. |
| 1979 | Suran, Jerome | 53 | Columbia | Mgr. |
| 1980 | Young, Leo | 54 | Ph.D. Johns Hopkins | Consult. |
| 1981 | Damon, Richard | 58 | Ph.D. Harvard '52 | Mgr. |
| 1982 | Larson, Robert | 44 | Ph.D. Stanford '64 | Mgr. |
| 1983 | Owens, James | 63 | Rice '41 | Consult. |
| 1984 | Gowen, Richard | 49 | Ph.D. Iowa State | Acad. |

# *The Societies of the IEEE*

| SOCIETY NUMBER | SOCIETY NAME | 1983 MEMBERSHIP |
|---|---|---|
| 1 | Acoustics, Speech, and Signal Processing | 8900 |
| 2 | Broadcast Technology | 3000 |
| 3 | Antennas and Propagation | 5400 |
| 4 | Circuits and Systems | 10 400 |
| 5 | Nuclear and Plasma Sciences | 2800 |
| 6 | Vehicular Technology | 2200 |
| 7 | Reliability | 3300 |
| 8 | Consumer Electronics | 5500 |
| 9 | Instrumentation and Measurement | 4800 |
| 10 | Aerospace and Electronic Systems | 6100 |
| 12 | Information Theory | 4300 |
| 13 | Industrial Electronics and Control Instrumentation | 4400 |
| 14 | Engineering Management | 7400 |
| 15 | Electron Devices | 8100 |
| 16 | Computer | 70 100 |
| 17 | Microwave Theory and Techniques | 6700 |
| 18 | Engineering in Medicine and Biology | 6900 |
| 19 | Communications | 16 500 |
| 20 | Sonics and Ultrasonics | 2100 |
| 21 | Components, Hybrids, and Manufacturing Technology | 2800 |
| 23 | Control Systems | 8700 |
| 25 | Education | 2100 |
| 26 | Professional Communication | 2200 |
| 27 | Electromagnetic Compatibility | 2100 |
| 28 | Systems, Man, and Cybernetics | 3900 |
| 29 | Geoscience and Remote Sensing | 1700 |
| 30 | Social Implications of Technology | 3400 |
| 31 | Power Engineering | 20 100 |
| 32 | Electrical Insulation | 1700 |
| 33 | Magnetics | 2700 |
| 34 | Industry Applications | 10 600 |
| 36 | Quantum Electronics and Applications | 3200 |

# *AIEE–IRE*
# *Merger Committee*

## AIEE REPRESENTATIVES

### *Chase, Warren H.*
(Cochairman) (President AIEE 1961–62)
AIEE F'51; IRE SM'51
Vice President, Ohio Bell Telephone Company, Cleveland, OH

### *Blackmon, Hendley*
AIEE F'49; IRE SM'57
Engineering Manager, Westinghouse Central Laboratories, Pittsburgh, PA

### *Clark, W. Russell*
AIEE F'61; IRE SM'47
Manager, Product Engineering Division, Leeds and Northrup Co.,
Philadelphia, PA

### *Linder, Clarence H.*
(President AIEE 1960–61)
AIEE F'57
Vice President and Group Executive, General Electric Company,
New York, NY

### *Robertson, Elgin B.*
(President AIEE 1953–54)
AIEE F'45
President, Elgin B. Robertson, Inc., Dallas, TX

### Robertson, Lawrence M.
AIEE F'45
Engineering Manager, Public Service Co. of Colorado, Denver, CO

### Teare, B. Richard, Jr.
(President AIEE 1962–63)
AIEE F'42; IRE F'51
Dean, Engineering and Science, Carnegie Institute of Technology,
Pittsburgh, PA

---

## IRE REPRESENTATIVES

### Haggerty, Patrick E.
(Cochairman) (President IRE 1962)
AIEE M'61; IRE F'58
President, Texas Instruments, Inc., Dallas, TX

### Berkner, Lloyd V.
(President IRE 1961)
AIEE F'47; IRE F'47
President, Graduate Research Center of the Southwest, Dallas, TX

### Henderson, John T.
(President IRE 1957)
AIEE M'57; IRE F'51
Principal Research Officer, National Research Council, Ottawa, Canada

### McFarlan, Ronald L.
(President IRE 1960)
IRE F'61
Consultant, Chestnut Hill, MA

### Peterson, Walter E.
AIEE AM'46; IRE SM'50
President Automation Development Corp., Culver City, CA

### Pratt, Haraden
(President IRE 1938)
AIEE F'37; IRE F'29
Consultant, Pompano Beach, FL

### Ryder, John D.
(President IRE 1955)
AIEE F'51; IRE F'52
Dean of Engineering, Michigan State University, E. Lansing, MI

---

## NONMEMBER SECRETARY

### Fink, Donald G.
(President IRE 1958)
AIEE F'51; IRE F'47
Meadowbrook, PA

# INDEX

Accreditation Board for Engineering and Technology, 204
Ada, Augusta, 179
Advertising for broadcasting, 75
Affel, H. A., 84
Aiken, Howard H., 8, 181
AIEE
  badge, 44
  Committee on Local Organization, 61
  Committee on Planning and Coordination, 219
  Committee on Units and Standards, 43
  early history of, 28, 33–34, 44–45
  and education, 194–197
  EE DIGEST, 219
  ELECTRICAL ENGINEERING, 227–228
  High Tension Transmission Committee, 61
  Institute Technical Groups, 220
  and IRE founding, 210
  membership policy, 64–65, 214–216, 219–220
  professionalism in, 63–64
  publication policy, 210
  Power Division and the merger, 209
  sections, 61
  and standards, 43–44
  students in, 61, 217
  Technical Committees, 61, 217
  technical organization, 210–212
  Telegraph and Telephone Committee, 61
  and television, 170, 171
  TRANSACTIONS, 40–41, 212
AIME, 45
  and Engineering Societies Building, 62
  professionalism in, 63
Alexanderson, E. F. W., 7
  alternator of, 51, 69–70
  television of, 154
  and triode, 53
Allgemeine Elektrizitats-Gesselschaft

(AEG), 97
Alternating current
  and direct current, 35–37
  early, 6–7
  frequency of, 39–40
  polyphase, 37–39
  in radio receivers, 79
Alvarez, L. W., 138
American Broadcasting Company, 80
American Marconi Company, 69–70
American Radio Relay League, 76
American Society for Engineering Education, 196, 204
American Standard Code for Information Interchange (ASCII), 175
Ampère, Andre-Marie, 2
Analog-to-digital, 187–188
Anthony, William A., 194
Antimony sulfide, 151
Appleton, Edward, 135
Arc light, 20
Armstrong, Edwin H., 133
  death of, 82
  and FM, 80–83
  on radio receivers, 79
  and RCA, 77, 81–82
  regenerative circuit of, 58–59, 74
  superheterodyne circuit of, 60–61
  and superregeneration, 77–78
  and Westinghouse, 70
Army Signal Corps Laboratories, 145
Arnold, B. J., 193
Arnold, H. D., 7
  high-vacuum tubes of, 56
  and Langmuir patent, 75
  in RCA, 70
  and triode, 59
Artificial intelligence, 188–189
Asheville Resolution, 219
ASCE, 45, 62–63
ASME, 45, 62

Atanasoff, J. V., 8, 182
Atari Company, 188
AT&T, *see also* Bell Laboratories
  and broadcasting, 75
  and RCA, 70–72, 75, 80
  and regeneration, 59
  TELSTAR, 146
Audion, *see* Triode
Automatic Sequence Controlled Calculator, 181

Babbage, Charles, 8, 178–180
Baer, Ralph, 188
Bailey, George W., 215, 224
Bain, Alexander, 150
Baird, John Logie, 153–156, 164
Baker, W. R. G., 81, 215
  and NTSC, 163, 165
Bardeen, John, 8, 121, 123, 129
Barker, George, 20
Barrett, Walter, 135, 210
Bartlett, Thomas W., 226
Becquerel, Alexandre, 7
Bell, Alexander Graham, 7, 16–20, 28, 31
Bell Laboratories, *see also* AT&T
  cavity magnetron, 138
  early analog computer, 86
  and FM, 82
  and information theory, 176
  integrated circuits, 128
  negative feedback, 83–84
  organization of, 57
  television, 153
  and transistor, 120–124
Bell Telephone Company, research at, 54–58
Benzer, Seymour, 124
Berg, Charles A., 113
Berkner, Lloyd, 221–222
Berliner, Emile, 86
Berliner Elektricitats-Werke (BEW), 97
Bits, definition of, 174
Black, Harold S., 83–84
Bohr, Niels, 8, 151
Bonneville Power Administration, 110–111
Boole, George, 8, 176
Boone, G. W., 187
Bradley, William, 126
Brattain, Walter, 8, 121, 123–124
Braun, Ferdinand, 8, 155
Braun tube, 155–156, 159
Bray, Ralph, 124
Breit, Gregory, 135
British Broadcasting Corporation, 155, 168
British General Electric Company, 137
Brown, Harold P., 37
Brown, Boveri Company, 39, 104

Brush, Charles F., 27, 29, 31, 91
  arc light of, 20
  at the 1884 Exhibition, 33
  and storage batteries, 35
Buck, H. W., 101
Bushnell, Nolan, 188
Bush, Vannevar, 140, 215
  and NDRC, 138
  network analyzer of, 102

Cables, telegraphic, 14–15
Caldwell, O. H., 117
California Electric Light Company, 92
Campbell, George, 54–55
Campbell-Swinton, A. A., 156–158
Canada, utilities in, 107–108
Carey, George R., 149
Carnegie, Andrew, 62–63
Carnegie Foundation, 203
Carnegie Institute, 135
Carson, John R., 57, 80–81, 83
Carty, J. J., 56, 57
"Cat's whisker," 8
Cavendish, Henry, 1, 3
Cavity magnetron, 137–140
Central stations, 92–94
  ac versus dc, 35–37
  first, 23–24
Chain Home radar, 136
Chamberlain, Joseph, 95
Chicago Edison Company, 100, 104–106
Christiansen, Donald, 229
Churchill, Winston, 137–138
Clarke, Arthur C., 145
Coal, 112–113
Codes, 174–175
Coffin, Charles A., 42, 104, 106
Coleco, 188
Colpitts, Edwin, 54, 56
Columbia Broadcasting System
  and color television, 164–165, 167–168
  founding of, 80
  and LP record, 86, 87
Commonwealth Edison Company, 104
Compton, Karl T., 138
Computers
  artificial intelligence, 188–189
  at Bell Labs, 86
  definition of, 177
  early, 86, 178–180, 182–183
  games, 188
  miniaturization of, 183–184
  programming, 186
Cooke, Morris, 108
Cooke, William, 11–13
Coolidge, Calvin, 110
Coolidge, W. D., 52–53

Cooper, Hugh L., 108
Cornell University, 194
Corona discharge, 101
Coulomb, Charles, 1
COLOSSUS, 133, 182
Craven, T. A. M., 162
Crone, W. Reed, 228
Crosley, 75
Cross, Charles R., 28, 30, 194
Curtis, Charles, 99–100
Cutler, C. Chapin, 228

Darlington, Sidney, 128
Davenport, Thomas, 6
Davy, Humphry, 20
deForest, Lee, 7
    on advertising, 75
    and regenerative circuit, 58–59, 70
    triode of, 51–52
de Laval, Carl Gustaf, 99
Deloy, Leon, 76
Dewitt, John H., Jr., 145
DeWitt, David, 228
Dieckmann, Max, 156
Digital recorders, 88
Digital-to-analog, 187–188
Diode, 48
    solid-state, 118–119
Dippy, Robert J., 142
Direct current
    and alternating current, 35–37
    Edison's system, 25
    high-voltage transmission of, 102
Dolby, R. M., 87
Dubridge, Lee, 138
Dumont Laboratories, 160
Duquesne Light Company, 113
Dushman, Saul, 61

EARLY BIRD, 146
Eastham, Melville, 142
Eckert, J. Presper, Jr., 8, 182–183
Eckert, W. H., 29, 30
ECHO, 85, 145
Edison, Thomas A., 28, 31
    and ac, 36–37
    at the 1884 Exhibition, 32
    generators of, 22–23
    Holborn Viaduct station of, 92–93
    incandescent lamp of, 6, 20–22, 91
    Pearl Street station, 93–94
    and phonograph, 85
    spread of system of, 27
    and storage batteries, 35
Edison effect, 7, 31, 34, 38
EDVAC, 183
Einstein, Albert, 8, 151

Electric Bond & Share Company, 106–107
Electric Power Research Institute, 114–115
ELECTRICAL ENGINEERING, 212
Electrical engineering education
    design courses in, 197
    faculty, 206
    graduate study, 200–202
    1900–1935, 198–200
    nonuniversity, 204–205
    origins, 194
Electrocution, 37
Electronics, 118
Electronics, origin of term, 117–118
Electronics Industry Association, 170
Elster, Julius, 151
Emberson, Richard M., 224
Emmett, W. L. R., 99
Engineering Council for Professional Development, 203–204, 214
Engineering schools, early, 192–193
Engineering, Science and Management War Training, 201
Engineering societies, organization of, 44–45
Engineering Societies Building, 62–63
England
    central stations in, 92–97
    early radio in, 79
    Electric Lighting Act, 94–95
    radar development in, 136–138
    utilities in, 108
ENIAC, 133, 182
ENIGMA, 181–182
Espenscheid, Lloyd, 84
Europe
    nuclear power in, 113
    television in, 169
Evans, Llewellyn, 111
Everitt, W. L., 206, 215

Fairchild Instruments, 127, 128
Faraday, Michael, 3–4, 6
    on conduction, 7
    and Cooke, 13
Farmer, Moses G., 6
Farnsworth, Philo T., 156, 160–161
Federal Communications Commission, 161–162, 165, 166
Federal government
    and electric utilities, 107–112
    radio licensing by, 60
    on radio ownership, 70–71
Federal Power Commission, 107
Ferguson, Louis, 104
Ferranti, Sebastian Z., 96, 99
Fessenden, Reginald A., 7
    and Alexanderson, 51
    and broadcasting, 74

heterodyne circuit of, 60
  and regeneration, 58
Field, Cyrus, 14–15
Field, S. D., 29, 31
Fink, Donald G., 224, 226, 230
  and IEEE Spectrum, 227–229
  meeting with Hickernell, 221
Fisk Street Station, 104–105
Fleming, J. A., 7, 48
Flynn, Ralph, 228
Fly, James L., 163
FM, 80–83
  mobile communications, 143–144
  stereo, 87
Forbes, George, 39
Ford, Henry, 110
Foster, W. J., 37
France
  television in, 169
  utilities in, 108
Franklin, Benjamin, 1
Franklin Institute, 1, 22, 31, 33–34
Freed–Eisemann Radio Corporation, 76
Frequency of ac, 39–40
Frequency modulation, see FM

Gallium arsenide, 127
Galvani, Luigi, 2
Gannett, Elwood K., 226, 228
Gaulard, Lucien, 35
GEE system, 142
Geitel, Hans, 151
General Electric Company
  and Alexanderson alternator, 51
  and American Marconi, 70
  and broadcasting, 75
  and FM, 81
  formation of, 38
  loudspeaker design, 86
  nuclear power, 113
  and RCA, 70–72, 75, 80
  research at, 52–54, 97
  silicon controlled rectifier, 130–131
  sound movies, 88–89
  steam turbines, 99–100, 104
  and Steinmetz, 42
  and Westinghouse, 103
Generators
  for arc lights, 22
  capacity, 100
  Edison's, 22–23
  Gramme's, 6, 20, 91
Gerber, Walter E., 169
Germanium, 8, 119–120
  carriers in, 122
Germany
  central stations in, 94, 97

high-voltage transmission in, 102
Getting, Ivan, 139
G. I. Bill, 201
Giant Power, 108
Gibbs, John D., 35
Glennan, T. Keith, 141
Goldmark, Peter, 86, 153, 164–165
Goldsmith, Alfred, 73–74
  and IRE–AIEE merger, 210
  and IRE Proceedings, 67, 227
  and organization of IRE, 66
  on professionalism, 229
  and radiola phonograph, 86
Goldstine, Adele, 179
Gowen, Richard J., 231
Gramme, Zenobe T., 6, 20
Gramophone, 86
Gray, Elisha, 16–18, 29, 31
Gray, Frank, 166–167
Gray, Truman, 118
Green, Norvin, 28
Grinter, L. E., 204

Haggerty, Patrick, 221–222, 224
Hall, Edwin, 8
Hamburger, Ferdinand, Jr., 227
Hamilton, George, 28, 30
Hammond, H. P., 204
Hammond, John Hays, Jr., 53
Hammond Electric Light Company, 93
Handie-Talkie, 144
Harding, Warren G., 110
Harrison Street station, 104
Hartley, R. V. L., 57, 76, 176
Harvard Radio Research Laboratory, 139
Haskins, Charles H., 28
Hazard, Rowland R., 34
Hazeltine Corporation, 167
Hazeltine, Alan, 76, 79
Heaviside, Oliver, 47, 54, 135
Heising, R. A., 57
Hellings, M. L., 29
Henney, Keith, 117–118
Henry, Joseph, 11
  and Bell, 18
  discovery of induction by, 4
  and early telegraphy, 13, 15
  and Gray, 18
  and Morse, 13
  and Wheatstone, 13
Herold, E. W., 167
Hertz, Heinrich, 6, 135
Herz, Eric, 224
Heterodyne circuit, 60
Hewitt, Peter Cooper, 104
Hewlett, E. M., 101
Hewlett-Packard, 128

Hibshman, Nelson, 223, 224
Hickernell, L. F., 219, 221
High-voltage transmission, 100–102
Hoerni, John, 127
Hoff, M. E., 187
Hogan, John V.
  and organization of IRE, 66
  tuning patent of, 76
Hollerith, Herman, 180
Hooven, Morris, 220
Hoover, Herbert, 110
Hopkinson, Edward, 23
Hopkinson, John, 23, 105
Hopper, Grace, 179, 215
Houston, Edwin J., 29, 31, 33–34
Hoxie, C. A., 88
Hubbard, Gardiner, 18
Hull, A. W., 61, 79
Hunter, T. A., 218–219
Hydroelectric power, 108–109
  first, 23
Hysteresis, 23
  Steinmetz on, 42

Iconoscope, 157, 159
IEE, 44
IEEE
  and ABET, 204
  badge, 223
  Committee on Professional Relations, 229
  Educational Activities Board, 205
  General Managers, 224
  headquarters, 225
  legal organization, 225
  Medal of Honor, 129
  name, 222–223
  POTENTIALS, 229
  and professionalism, 229–231
  size, 209
  Service Center, 226–227
  Society on the Social Implications of Technology, 231
  SPECTRUM, 227–229
  STUDENT JOURNAL, 229
  and television, 170, 171
  U.S. Activities Board, 230
ILLIAC, 183
Incandescent lamp, 6, 20–25
  Edison's, 91
  German, 97
  improved filaments for, 52–53
  Swan's, 92
Industrial research
  at Bell Telephone, 54–58
  at G.E., 52–54
Information
  definition of, 173–174

transmission of, 175–177
Institution of Civil Engineers, 44
Insull, Samuel, 93, 103–106
Integrated circuits, 8, 127–130
Intel Corporation, 129, 187
INTELSAT, 146
International Business Machines Corporation, 180, 183
International Electrical Exhibition, 28, 31–33
International Electrical Congress, 43
International Telecommunications Satellite Consortium, 146
IRE
  and Armstrong, 59
  badge, 66
  early autonomy, 210
  headquarters, 225–226
  membership policy, 67, 214–216, 218–219
  Organization of, 65–67
  PROCEEDINGS, 67, 212–214, 227–228
  Professional Technical Groups, 216–217
  publication policy, 212–214
  on radio ownership, 70
  Sarnoff in, 73
  sections, 67
  Spring Convention, 214
  STUDENT QUARTERLY, 216, 218–219
  students in, 217
  technical organization, 210–212
  and television, 170, 171
  TRANSACTIONS, 217
Itaipu, 109
Italy, utilities in, 108
Ives, Herbert E., 154, 164

Jablochkoff candle, 91
Jackson, Dugald C., 194
Jacquard, J. M., 178
Jansky, Karl, 57–58
Japan, television in, 169–170
The Jazz Singer, 88
Jeffries, Zay, 118
Jenkins, Charles F., 153
Jewett, Frank B., 55–57
Johnson, Harwick, 128
Joint Technical Advisory Committee, 165
Jones, F. W., 29, 30
Jones, Thomas F., Jr., 227
Joule, James, 5–6

KDKA, 74, 75
Keith, N. S., 28
Kellogg, E. W., 86
Kelly, Mervin, 121

Kennelly, Arthur E., 41–43, 47, 135
  and AIEE–IRE merger, 210
  and Giant Power, 108
Keokuk, Iowa, 108
Kilby, Jack, 8, 128
Kinescope, 159
King, Donald D., 231

Lamme, Benjamin G.
  and induction motor, 94
  and Steinmetz, 53
  and Tesla, 37
Langmuir, Irving, 7, 56
  and Arnold patents, 75
  and RCA, 70
  and triode, 53, 59
  and vacuum tube nomenclature, 61
Lark-Horovitz, Karl, 119–120, 124, 133
Lasers, 85, 131
Lauffen–Frankfort line, 94, 101
Law, H. B., 167
Lead sulfide, 8
Leibniz, Gottfried, 176
Lewis Task Force, 220
Lewis, W. A., 220
Lightning and transmission lines, 102
Lilienfeld, J. E., 121
Linder, Clarence, 221, 222, 225
Load diversity, 105–106
Load factor, 105–106
Lodge, Oliver, 7
  coherer of, 47
  tuning patent of, 48, 69
London Electric Supply Company, 96
Loomis, Alfred, 138
Loran, 142–143
Loughlin, Bernard, 167
Lovell, C. A., 86
Lowenstein, Fritz
  and organization of IRE, 66
  and triode, 53, 56
  tuning patent of, 76
Lowrey, Grosvenor, 21

Magnavox, 188
Magnetic circuit, 23
Marconi, Guglielmo, 7, 47–48
  and broadcast frequency, 59–60, 76–77
Marconi–EMI television, 155
Marriott, Robert
  and organization of IRE, 66
  on Sarnoff, 72
Mauchly, John W., 8, 182–183
Maxwell, James Clerk, 4, 6
May, Joseph, 150
McCue, J. J. G., 228
McFarlan, Ronald, 221

Meissner, Alexander, 58
Mercury-arc rectifier, 104
Merger Committee
  and IRE finances, 214
  and ITG's, 220
  organization, 221–222
Mershon, Ralph, 64, 101
Merz, Charles, 96
Merz, J. Theodore, 96
Microprocessors, 187
Microwaves, 85
Millikan, Robert, 7, 8
Mills, John, 118
Mitchell, S. Z., 106–107
MKSA system, 44
Mobile communications, 143–144
Moore, D. McFarlan, 151
Moore, Gordon, 129
Morgan, Stanley, 121
Morrill Land Grant Act, 192–193
Morse, Samuel F. B., 7, 13–14
Motorola, 82
  mesa transistor, 127
  mobile communications, 144
Motors
  ac polyphase, 37, 38
  early, 6
  induction, 7, 37, 38, 94
  for transit, 98
Movie sound, 88–89
Muscle Shoals, 108, 109
M.I.T., 194
M.I.T. Radiation Laboratory, 119, 138–140

National Broadcasting Company, 80
National Conference of Electricians, 31, 33
National Defense Research Committee,
  138–139
  and loran, 142
National Electric Signaling Company, 51
National Television Systems Committee,
  163–170
Negative feedback, 83–84
Nernst, Walter, 97
Network analyzers, 102–103
Neutrodyne, 75–76, 78–79
New Brunswick, NJ, American Marconi at,
  69–70
Newcastle on Tyne Electric Supply Com-
  pany, 96
Niagara Falls, 39–40, 94, 108
Nicholson, Alexander M., 158, 159
Nipkow disk, 151–154, 157
Noble, Daniel, 82, 143–144
Norris, George W., 109–110
Noyce, Robert, 8, 128–129
N-type semiconductors, 122

Nuclear power generation, 113
  in England, 108
Nyquist, Harry, 84

Oerlikon Company, 38, 39, 97
Oersted, Hans, 2
Ohm, Georg Simon, 2–3
Oliver, Bernard, 128
  and IEEE badge, 223
ORDVAC, 183
Orthicon, 159
Orthophonic phonograph, 86
Owens, R. B., 194

Pacific Intertie, 102
Pacinotti, Antonio, 6
Paiva, Adriano de, 150
Parkinson, D. B., 86
Parsons, Charles, 99
Pearl Street station, 23–24
Peek, F. W., Jr., 102
Petroleum, 113
Philco Corporation, 126, 127
Phonograph, 85–87
Pierce, John A., 142
Pierce, John R., 227
Pitts, Bill, 188
Planck, Max, 8, 151
Pope, Frank L., 28–30
Pope, Ralph, 210
Popov, Alexander, 48
Poulsen, Valdemar, 87
Preece, William, 34
Prescott, George B., 29
P-type semiconductors, 122
Public Service Company of Northern Illinois,
  106
Pumped storage stations, 100
Pupin, Michael, 54–55
Pupin, Michael I., 41, 43, 54–55
  loading coils of, 175
Purcell, E. M., 138

Quarry Street station, 105

Rabi, I. I., 138
Radar
  peacetime applications, 140
  pre-World War II, 134–137
  during World War II, 137–140
Radiation Laboratory series, 140
Radio
  advertising for, 75
  amateurs, 59, 76–77
  broadcast frequency of, 59–60
  domestic receivers, 78–79

earliest, 47–51
early broadcasting, 74–75
licensing of, 60
mobile, 82
sidebands, 83
in World War I, 60
Radio astronomy, 58
Radio Manufacturer's Association, 154,
  161–163, 170
Radio Research Laboratory, 139
Radio waves, discovery of, 6
RCA
  antitrust action against, 80
  and broadcasting, 74
  color television, 165, 166–168
  and Farnsworth, 160–161
  and FM, 81–82
  and 45 rpm record, 87
  integrated circuits, 128
  location of first laboratory, 74
  manufacturing by, 74–75
  and Neutrodyne, 77
  organization of, 70–72
  sound movies, 89
  superheterodyne, use of, 79
  and phonograph, 86, 87
  and superregeneration, 77
  and television standards, 161–163
  vidicon camera tube, 151
  and Zworykin, 159
Reber, Grote, 58
Redmond, Denis D., 149
Regenerative circuit, 58–59, 70–71
Reinartz, John, 76
Reist, H. G., 37
Remington–Rand, 183
Rensselaer Institute, 192
Rice, C. W., 76, 86
Rice, E. W., 42
Roosevelt, Franklin D., 110, 111
Rosing, Boris, 156
Rotary converter, 103
Round, H. J., 58
Rowland, Henry, 8, 23
  at the 1884 Conference, 33
Rubidium in television, 156
Rural Electrification Administration,
  110–111
Russell, Steve, 188
Russia: hydroelectric power in, 108
Ryan, Harris J., 101
Ryder, John D., 223
  and IEEE SPECTRUM, 227–228
  meeting with Hooven, 220

Sammet, Jean, 179
Samuel, Arthur, 188

Sanabria, Ulises A., 157
Sargent, Frank, 104
Sarnoff, David
  and color television, 165, 168
  early years, 72
  and FM, 81
  and NBC, 80
  and superregeneration, 77
  and television, 160, 162
Satellites, 144–147
Schnell, Fred, 76
Schottky, Walter, 79
Schroeder, A. C., 167
Schulke, Herbert A., 224
Scott, Charles F., 193
  as AIEE President, 61–63
  high-voltage studies of, 101
  and induction motor, 94
  T connection of, 39, 40
  and Tesla, 37
SCR 268, 136
SCR 584, 139
Selenium, 8, 150
Semiconductors, discovery and explanation
  of, 7–8
Senlecq, Constantin, 150
Shallenberger, O. B., 37
Shannon, Claude, 8, 133, 176–177
Shippingport, PA, 113
Shockley, William, 8, 121, 126, 128–129,
  215
Siemens and Halske, incandescent lamp of,
  97
Siemens, Werner von, 6, 97
Silicon, 8, 119–120, 122, 127
Silicon controlled rectifier, 130–131
Single-sideband transmission, 83
Sirjane, Emily, 226
Smith, David, 126
Smith, Willoughby, 8, 150
Smith, W. W., 29
Society for the Promotion of Engineering
  Education, 196, 203–204
Society of Telegraph Engineers, 44–45
Society of Wireless Telegraph Engineers, 66
Sonar, 141–142
Space exploration, 144–145
  communication in, 177
Sporn, Philip, 113
Sprague, Frank, 32, 98
Sputnik, 144
Stanley, William, 6, 36
State Line station, 100
Steam engine, 99
Steam turbine, 40, 99–100
  Chicago Edison, 104–105
Stearns, Joseph, 16, 17

Steinmetz, Charles P., 23, 41–43
  on education, 197
  and German lamps, 97
  and Lamme, 53
  and research, 52
Stereo, 87
Stern, Arthur P., 230
Stillwell, L. B., 40
Stoney, G. Johnstone, 7, 48
Stone, John Stone, 54, 56, 66
Storage batteries, 35
Strowger, Almon B., 57
Superheterodyne circuit, 60–61, 79
  in England, 79
  Westinghouse purchase of, 70
Superregeneration, 77–78
Swan, Joseph, 6, 20, 21, 33, 92
SYNCOM, 146

Tape recorders, 87–88
Taylor, A. Hoyt, 135–136
Teare, B. R., 221–222
Telegraph, 7, 11–16
  code for, 174–175
Telephone, 7, 16–20
  information transmission by, 175
  long distance, 54–57, 84–85
  switching, 57
  vacuum tube as a repeater, 121
Television
  color, 164–168
  compatibility, 73, 164–165
  early, 149–158
  electronic, 158–159
  information transmission by, 175
  international, 168–170
  practical, 160–164
  satellite transmission of, 146–147
  standards, 161–163
  UHF, and radar, 140
Telluride, Colorado, 101, 108
TELSTAR, 85, 146
Tennessee Electric Power Company, 111
Tennessee Valley Authority (TVA), 110–112
Terman, F. E., 139
  and IRE organization, 215
  *Radio Engineering*, 206
  research productivity report, 199–200
  and Silicon Valley, 201
Tesla, Nikola, 7
  induction motor of, 94
  and Westinghouse, 37, 38
Tetrode, 79
Texas Instruments Corporation, 122
  integrated circuits, 128
  silicon transistor, 127
Thayer, Sylvanus, 192

Thompson, Silvanus P., 22
Thomson, Elihu, 30
  and ac equipment, 38
  at the 1884 Exhibition, 32–33
  and induction motor, 94
  and Steinmetz, 42
Thomson, J. J., 7, 48
Thomson, William, 14–15
Thomson–Houston Electric Company, 30
  and ac equipment, 38
  and ac motors, 37
  at the 1884 Exhibition, 32–33
Three Mile Island, 113
Transformers, first, 35
Transistors
  diffused-base, 126
  field-effect, 121
  junction, 8, 121, 124–126
  mesa, 127
  micro-alloy, 126
  planar, 127
  point-contact, 8, 121–124
  silicon, 127
  surface-barrier, 126
Transit, 97–98
Triode, 51–52
  in early receivers, 78–79
  G.E. research on, 53
  patent case, 75
  in RCA, 70
  as telephone repeater, 56
Trowbridge, W. P., 29, 30
Turing, Alan, 182
Tuve, Merle, 135
Tykociner, J. T., 88

United Engineering Trustees, 62
United States Electric Lighting Co., 32
United States Military Academy, 192
UNIVAC, 183
University of Missouri, 194
University of Wisconsin, 194
Upton, Francis, 21
U.S. Naval Research Laboratory, 136
U.S. Navy Underwater Sound Laboratory, 141
U.S. Signal Corps, 136

Vacuum tubes
  in ENIAC, 182
  in IBM 650, 183
  modern use of, 118
  weaknesses of, 120–121
Vail, Alfred, 7, 14
  code of, 174
Vail, Theodore N., 29, 56, 57
Van Depoele, Charles, 98
Volta, Alessandro, 2

von Braun, Werner, 144
von Neumann, John, 8, 133, 182–183

Walker, Eric, 204
Walkie-Talkie, 144
Watson, Thomas J., 183
Watson-Watt, Robert, 136
Watt, James, 4–5
WEAF, 75, 80
Weber, Ernst, 225
Weiner, Norbert, 133
Welsbach, Carl, 97
Welsbach mantle, 95
Western Electric
  and phonograph, 86
  sound movies, 88–89
  and television, 157
Westinghouse, George, 35–37
Westinghouse Electric Company
  and broadcasting, 74–75
  at Niagara Falls, 39–40
  nuclear power, 113
  patent exchange with G.E., 103
  and polyphase ac, 37–39
  in radio, 70–71
  and RCA, 71, 75, 80
  steam turbines, 99–100
  and Zworykin, 159
Weston, Edward, 22, 29, 31
  at the 1884 Exhibition, 32
  motors of, 27
WGY, 75
Wheatstone, Charles, 13
  telegraph of, 7
Whitney, Willis R., 52, 53
Wickenden, W. E., 203
Willenbrock, F. Karl, 228
Wilson Dam, 109
Wilson, Woodrow, 70
Wireless Institute, 66
WJZ, 75
WLW, 75
World War I and electrical engineering
  education, 199
World War II
  and electrical engineering education, 201–203
  electronics' importance in, 118
  and television, 163–164
  and TVA, 111–112
Wright, Arthur, 93
  and Insull, 105
WWJ, 74

Yankee Network, 81
Yokogawa Electric, 103
York, R. A., 130

Zworykin, Vladimir K., 156–161

# PICTURE CREDITS

# AUTHORS' BIOGRAPHIES

## *John D. Ryder*

John D. Ryder received the B.E.E. and M.S.E.E. degrees from Ohio State University in 1928 and 1929. He was employed for two years by the General Electric Company in vacuum and gas tube development. From 1931 to 1941, he was in charge of industrial electronic applications for Bailey Controls and received 24 patents covering small-motor control and temperature measurement.

In 1941 he went to Iowa State University as an Assistant Professor of Electrical Engineering, received the doctorate there in 1944, and advanced to Professor. He helped design the Iowa State High-Frequency Network Analyzer (10 000 Hz), which led to close relations between that department and the electrical utilities. He was appointed Assistant Director of the Iowa Engineering Experiment Station in 1947.

He became head of the Department of Electrical Engineering, University of Illinois, in 1949, and in 1954 was made Dean of the College of Engineering, Michigan State University. He fostered miniaturization and modeling in the laboratories, and helped to achieve several curricular reforms. In 1966–68 he served as Vice Chief of the USAID Higher Education Project in Brazil. Upon his return, he assumed the rank of Professor of Electrical Engineering at Michigan State. He moved to Florida in 1972 and has taught there at the University of Florida.

He has authored seven textbooks in electronics and circuit theory, as well as many technical papers.

He was elected to the Board of Directors of the IRE in 1952, was President of that organization in 1955, and Editor in 1958 and 1959. He was a member of the 14-man Merger Committee to form the IEEE, and was the first Editor of the IEEE as well as a member of the first Board of Directors. SPECTRUM magazine was launched under his direction.

Dr. Ryder has been chairman of the IRE and AIEE Education Committees, the IEEE Fellow Award Committee, and the IEEE History Committee. In 1974 he was IEEE Executive Vice President. A Fellow of the IEEE, he received the Haraden Pratt Award of the Institute in 1979. He is currently chairman of the Task Force for the 1984 IEEE Centennial program.

# Donald G. Fink

Donald G. Fink received the B.Sc. degree from M.I.T. in 1933 and the M.Sc. degree from Columbia in 1942, both in electrical engineering. After serving as a Research Assistant in the Departments of Electrical Engineering and Geology at M.I.T., he was on the staff of McGraw-Hill's *Electronics* magazine from 1934 to 1952, holding the position of Editor from 1946 to 1952. On leave from McGraw-Hill, he was Head of the Loran Division, M.I.T. Radiation Laboratory, from 1941 to 1943 and an Expert Consultant, Office of the Secretary of War, from 1943 to 1946.

From 1952 to 1960, he was Director of Research for the Philco Corporation. He was appointed that company's Vice President for Research in 1960 and became Director of Philco-Ford Scientific Laboratories in 1962.

In 1963 he was selected as the first General Manager of the IEEE, which was formed that year from a merger of the AIEE and the IRE. He continued as IEEE General Manager and Executive Director until 1974, at which time he was appointed Director Emeritus for life. Since leaving the IEEE staff, he has carried out a variety of engineering consulting and editorial assignments.

Mr. Fink is the author of a large number of technical papers and holds two patents on aural stereophonic systems. He has written or edited 13 books, which span from the 1938 *Engineering Electronics* to the 1982 edition of *Electronics Engineers' Handbook*. He served as IRE Editor in 1956 and 1957 and as President in 1958.

Among his awards and honors are the following: Eta Kappa Nu Outstanding Young Engineer (1940), Medal of Freedom (1946), Presidential Certificate of Merit (1948), American Technologists Award (1958), membership in the National Academy of Engineering (1969), IEEE Founders Medal (1978), and Progress Medal of the Society of Motion Picture and Television Engineers (1979).

In 1979 the annual Donald G. Fink Prize Paper Award was established for the best tutorial, review, or survey paper published in any IEEE journal.

He is a Fellow of the IEEE, the Society of Motion Picture and Television Engineers, and the Radio Club of America. His other memberships include Sigma Xi, Eta Kappa Nu, and Tau Beta Pi.